S0-AKX-874

Stationary States

ALAN HOLDEN

Stationary States

1971
OXFORD UNIVERSITY PRESS
New York and Oxford

Library of Congress Catalogue Card Number: 75-167753
Printed in the United States of America

General Preface

THIS monograph is one of a connected series of three by the same author: *Stationary states*, *The nature of atoms*, and *Bonds between atoms*. Like its fellows, it was written for the Conference on New Instructional Materials in Physics, sponsored by the Commission on College Physics with the support of the National Science Foundation, and held at the University of Washington in the summer of 1965.

The purpose of the conference was to create effective ways of presenting physics to college students who are not preparing to become professional physicists. Such an audience might include prospective secondary-school physics teachers, prospective practitioners of other sciences, and those who wish to learn physics as one component of a liberal education.

The form of these, and of other monographs originating at that conference, implements a principle of experimental pedagogy promulgated there. Each monograph is 'multi-level': each consists of several sections arranged in order of increasing sophistication. Their authors hope that such papers can be fragmented for use in existing formal courses, or can be associated with other papers to provide the basis for new kinds of courses.

Contents

Foreword

THIS monograph is not intended to provide an introduction to quantum mechanics. It is a summary of the quantum-mechanical ideas and methods used in two monographs, *The nature of atoms* and *Bonds between atoms*, by the same author. Presumably you have met these ideas before. By riding over them rough-shod, in a cumulative order, the monograph offers a brisk review of them, and a helpful resort if you read the other two monographs.

The wave-mechanical formulation of quantum mechanics appearing here is that usually employed in discussing most of the properties of matter. It is somewhat easier to make mental pictures with this formulation than with others. Nevertheless much may seem strange, and much rather arbitrary, in quantum mechanics. Familiarity, and a sense of the beauty and consistency of the theory, will hardly come from reviewing it as briefly and fragmentarily as this monograph can do. But the application of its principles elsewhere will improve your feeling for the subject.

The monograph sketches only those aspects of quantum mechanics that are of most direct use in picturing and calculating the behaviour of atoms when they are stably bonded together in molecules and solids. Its emphasis is on stationary states; it does not deal with processes. The examples with which it illustrates its points are simplified by such great artificiality that they may seem irrelevant to the natural world. But in fact their simplicity does not remove their relevance; indeed it exemplifies the artificial simplicity usually necessary for arriving at a physical understanding of the complexities of the world.

ALAN HOLDEN

Stationary States

1. Wave-like and Particle-like Behaviour

ACCEPTING the idea to which Sir Ernest Rutherford's experiments led him in 1911—the idea that an atom consists of a positively charged nucleus and some negatively charged electrons, all much smaller than the atom that they build—the imagination constructs a picture. In that picture the relatively light electrons revolve about the relatively heavy nucleus in orbits, like the relatively light planets revolving about the relatively heavy sun.

Indeed the planetary model of an atom is very appealing. The attractive force obeys the same law of distance in both cases: it varies with the inverse square of the distance between the attracting centres. In the planetary case the attractive force is gravitational (proportional to the product of the two masses); in the atomic case the force is electrostatic (proportional to minus the product of the two charges), for a calculation using the known ratio of charge to mass of the electron quickly shows that the gravitational contribution in an atom can be neglected.

But observations of atomic behaviour bring to light two obstinate difficulties with this model. Call them Difficulty (a) and Difficulty (b).

Difficulty (a). A charged body, revolving about an oppositely charged body, should continuously radiate an electromagnetic wave, as a radio antenna does. An electromagnetic wave carries energy, the energy of the whole system is conserved, and thus the model continuously loses energy to the electromagnetic wave, until the electron falls into the nucleus. But real atoms are remarkably stable; matter is not constantly radiating electromagnetic energy in such a fashion.

Difficulty (b). The model provides a continuous spectrum of possible orbits for the electrons, permitting the total energy of the atom to have any value whatever. When the model is disturbed, by an incoming electromagnetic wave for example, it accepts any amount of energy that it gets a chance to accept, by making appropriate changes in the electronic orbits. But a real atom accepts energy only in certain definite *quanta*, whose size is closely connected with the frequency of the electromagnetic wave from which it absorbs the energy. And when excited to a condition in which it has more than its normal amount of energy, it will emit that energy in similar quanta. Indeed the frequencies associated with the wavelengths of the resulting spectral lines provide most of our quantitative information about the structure of atoms.

These difficulties are partly resolved by quantum mechanics. Only partly, for the newer mechanics still says little about what happens while an atom is changing from one of its energy states to another. But quantum mechanics provides a consistent description of the phenomena actually observed—the spectral lines—and the property immediately inferrable—the existence of stable states of fixed energy—and perhaps that is all that should be asked of any physical theory. Indeed, in its wave-mechanical formulation, quantum mechanics goes a little further and offers the beginnings of a picture of these happenings and inferences. The picture is ambiguous but it is the best available at present to assist a visualization.

The ambiguity in the wave-mechanical picture is often called *wave–particle duality*. Electrons, ordinarily thought of as particles, have some of the properties of waves; and light, ordinarily thought of as waves, has some of the properties of a stream of particles. Perhaps the wave-like character of electrons will seem more readily acceptable after examining the particle-like character of light.

The familiar wave theory of light starts out as successfully as the particle theory of the electron, taking such phenomena as refraction and diffraction in its stride. Then it too encounters two obstinate difficulties—Difficulty (c) and Difficulty (d).

Difficulty (*c*). Monochromatic light falling on a metal ejects electrons, and the kinetic energies of the ejected electrons range up to a maximum value which depends on the frequency of the light and on the species of the metal, but not on the intensity of the light. The total number of electrons ejected depends, of course, on the intensity of the light, and on how long it shines. But the maximum energy of an ejected electron, and the fraction of electrons having a particular energy, depend not at all on the duration and intensity of the light.

The wave theory of light can make no convincing picture of this behaviour. The light is supplying energy in proportion to its intensity and duration. But the ejected electrons do not show an energy greater than $E = h\nu - w$, where ν is the frequency of the light, w is characteristic of the metal, and h is Planck's constant, the fundamental quantity ubiquitous in quantum theory, with the value $6{\cdot}62 \times 10^{-34}$ Js. Moreover, experiments in which small bits of metal are illuminated have shown that such bits may eject electrons with energy E before the bits have had time to absorb that much energy according to the wave theory of light.

Difficulty (*d*). It is convenient to think that waves propagate in a *medium*. The imagined medium supports the waves; its undulations 'are' the waves. The medium propagates the waves with a velocity characteristic of the medium, and if you move through the medium you must be able to observe different apparent propagation velocities according to whether you move along the direction of propagation or across that direction (Fig. 1.1). But in the case of light, as Albert Michelson and Edward Morley found in 1887, the apparent velocity is always the same: $c = 3 \times 10^{8}$ m s^{-1}.

Difficulty (c) can be handled by supposing that light is a stream of particles, each of which has a kinetic energy $h\nu$, and which behave like a wave motion when they are observed in large numbers, but which individually perform little acts upon the atoms as if they were colliding with the atoms. The particles are often called *photons*.

Difficulty (d) is not removed by this supposition: The experiment diagrammed in Fig. 1.1 formed a foundation for the special theory of relativity. That theory starts with two premises: (1) Only the motion of one thing relative

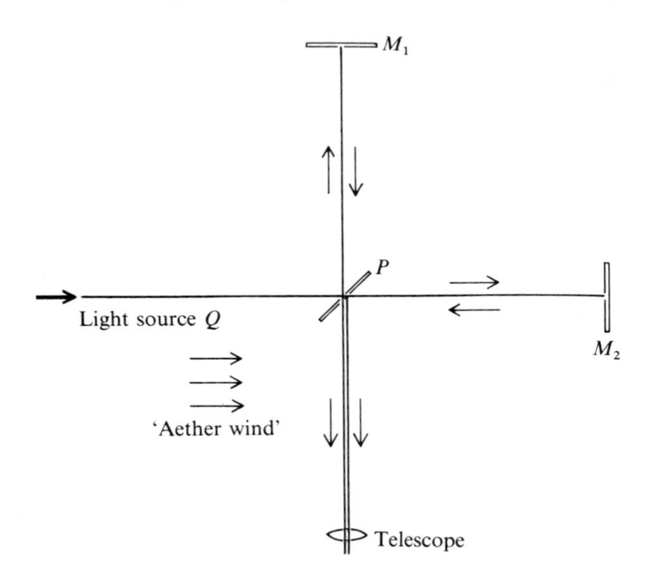

FIG. 1.1. Michelson and Morley divided light from Q into two beams by the half-silvered mirror P, reflected the beams from M_1 and M_2, and reunited them at P. The interference fringes observed in the telescope should be displaced when the apparatus is turned so that first M_1P and then M_2P are along the direction of the 'aether wind'.

to another, and not the absolute motion of something taken alone, has any meaning. (2) The velocity of light is finite, and all observers of it will obtain the same value, regardless of their relative motion. The consequences of these statements, derived by preserving the principles of conservation of momentum, and by welding the separate principles of the conservation of mass and of energy into a single principle of the conservation of mass-and-energy, provide a dynamical theory of particles called the *special theory of relativity*. The difference between this newer mechanics and the more familiar mechanics of Newton is signalized by the unique position of *the speed of light*, c, in the new theory. The predictions of the two theories depart further from each other the more nearly the speed of a particle approaches the speed of light.

It is therefore especially interesting to examine what the theory of relativity

Discussion 1.1

THE MOMENTA OF PHOTONS

This is not the place to develop the arguments of the theory of relativity. It is appropriate only to exhibit a few results of that theory and to show that they can lead to the relation between the momentum and the wavelength of a photon. In that theory it is found possible to retain the familiar expression for the *momentum*, p, of a particle,

$$p = mv, \tag{D1.1}$$

by using for its mass, m, an *apparent* mass,

$$m = \frac{m_0}{\sqrt{(1-v^2/c^2)}}, \tag{D1.2}$$

where m_0 is the *rest mass* of the particle and v is its velocity relative to the observer. The *total energy* of the particle is shown to be

$$U = c\sqrt{(p^2+m_0^2c^2)}. \tag{D1.3}$$

When the rest energy, m_0c^2, is subtracted from this expression, the remainder is the kinetic energy:

$$\begin{aligned} U_{\text{kin}} &= c\{\sqrt{(p^2+m_0^2c^2)}-m_0c\} \\ &= m_0c^2\left\{\frac{1}{\sqrt{(1-v^2/c^2)}}-1\right\} = c^2(m-m_0). \end{aligned} \tag{D1.4}$$

Accepting this solution offered by the theory of relativity to Difficulty (d), and also the solution to Difficulty (c) that light behaves like a stream of photons, each with kinetic energy $h\nu$, you can use the above relations to inquire into other properties of photons. Clearly these relations would give them infinite apparent masses, momenta, and kinetic energies unless their rest masses are set equal to zero. Using the value $m_0 = 0$, eqn (D1.4) yields for the kinetic energy of a photon

$$U_{\text{kin}} = mc^2 = h\nu. \tag{D1.5}$$

Hence $m = h\nu/c^2$, and eqn (D1.1) yields for the momentum of a photon

$$p = mc = h\nu/c = h/\lambda, \tag{D1.6}$$

where λ is the wavelength of the light.

says about the only particles known to travel at the speed of light, namely, the particles of light itself, the photons. As Discussion 1.1 points out, the theory would give to such particles infinite apparent masses, momenta, and kinetic energies unless their rest mass were zero. But since a photon is observed only when it is travelling at the speed of light and never when it is at rest, there is no contradiction with experiment in supposing that its hypothetical rest mass vanishes. Then the assumption that a photon has an energy $h\nu$ leads to the conclusion that its momentum is given by $p = h/\lambda$, where λ is the wavelength of the light.

This argument suggests a physical picture of light; it is a train of waves, and at the same time a stream of particles of zero rest mass, travelling at the speed of light. And the picture in turn suggests an experiment. When a photon collides with a material body at rest, it should impart some of its momentum and some of its kinetic energy to the body, losing some of its own. But since its momentum is connected with its wavelength, by $p = h/\lambda$, its wavelength should be increased and its frequency thus reduced by the collision.

In 1922 Arthur Compton reported experiments verifying this prediction. In *Compton scattering* (Fig. 1.2) X-rays scattered by nearly free electrons

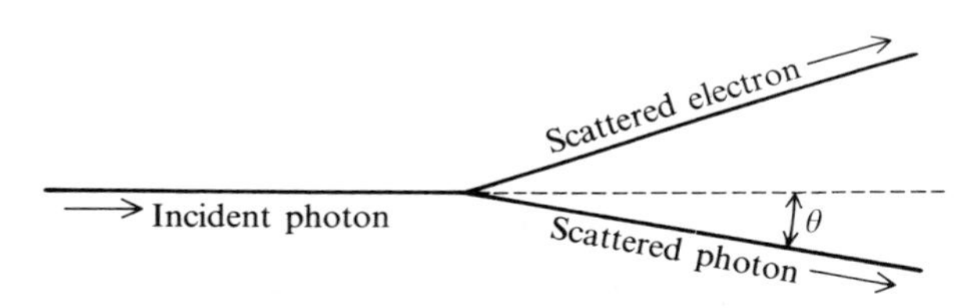

FIG. 1.2. In the Compton scattering of photons by electrons, the laws of conservation of energy and momentum, and the quantum-mechanical connection between the energy and the frequency of a photon, make it possible to predict the observed change in wavelength of the photon: $\Delta\lambda = (h/mc)(1 - \cos\theta)$.

show the predicted change in wavelength, as a function of the angle of scattering. Substituting the values of h, c, and the mass m of an electron into the relation shown in Fig. 1.2, and recalling that X-rays have wavelengths of the order of one ångström (10^{-10} m), you find that the change of wavelength can be as great as two per cent.

Wave mechanics constructs a wave-like picture for all particles, somewhat like the picture long accepted for photons. It gives an unexpected wave-like character to material particles, complementing the unexpected particle-like character of light. You can think of a moving particle as a travelling wave-packet, with a wavelength related to the momentum of the particle by a relation proposed by Louis de Broglie in 1924:

$$p = h/\lambda. \tag{1.1}$$

Notice that the *de Broglie relation* (1.1) is the same as the relation (D1.6) derived for photons in Discussion 1.1. But of course, material particles and their associated waves differ importantly from photons. The material particles do not have vanishing rest masses, and their waves are not electromagnetic. What the waves are, and how the picture of them resolves Difficulties (a) and (b), are the subjects of the next few chapters.

Again, however, the picture makes an immediate prediction: there should be some observable phenomenon in which material particles are diffracted like waves. The observation of such a phenomenon would require diffraction gratings with which particles can interact, and ways of imparting a uniform momentum to the particles so that they will have a monochromatic de Broglie wavelength in the right range for the grating to diffract them.

Crystals make good diffraction gratings for X-rays, and metal crystals should be especially good gratings to diffract electrons. Since metal crystals are electrical conductors, electrons can enter and leave them especially easily. Electrons emitted by a hot filament, and accelerated by a positively charged cylindrical casing around it, will escape through a slit in the casing, forming a beam of electrons with quite uniform velocities. The velocities appropriate for diffraction by a crystal are readily attainable, as Discussion 1.2 shows.

Discussion 1.2

THE WAVELENGTHS OF ELECTRONS

An electron accelerated from rest through a potential difference will acquire a kinetic energy $U_{\text{kin}} = \frac{1}{2}mv^2$, and a momentum $p = mv$, where m is its mass and v is its terminal velocity. The corrections offered by the theory of relativity can be neglected, so long as v, the velocity acquired by the electron, is small compared with c, the velocity of light. Then by the de Broglie relation $p = h/\lambda$, its wavelength will be $\lambda = h/\sqrt{(2mU_{\text{kin}})}$. Its kinetic energy will be equal to the potential difference times the charge on an electron, e. By using the values of h, m, and e, the wavelength can be expressed in terms of the voltage V applied to the electron gun: $\lambda = 12{\cdot}2/\sqrt{V}$ in ångströms. Since the spacings of atoms in crystals are a few ångströms, a few hundred volts on the gun gives the electron a wavelength suitable for it to be diffracted by a crystal.

In 1927 Clinton Davisson and Lester Germer in the United States, and George Thomson in England, found that such an electron beam will in fact afford diffraction patterns from the crystals in thin layers of metal, very similar to the patterns given by X-rays from thicker layers. The principal differences come largely from the higher scattering probabilities for electrons. Accelerated by potential differences of hundreds of volts, the electron beam is mostly diffracted by a few layers of atoms near the surface of the crystal. When potential differences of thousands of volts are applied, the electrons penetrate the crystal much further.

Thus a curious symmetry seems to inhere in nature, by which both wave-like and particle-like behaviour can be discerned in all of nature's fundamental processes. Often we can disregard this wave-particle duality, and describe a process by methods developed to describe exclusively the properties of waves or of particles as we have long understood them. But this is usually possible only in large-scale phenomena where multitudes of wave–particles are cooperating. In small-scale happenings we must remember that a *particle* behaves also like a *wave-packet*; and we must connect its particle-like behaviour, described by its momentum p, with its wave-like behaviour, described by its wavelength λ, through the de Broglie relation $p = h/\lambda$.

PROBLEMS

1.1 By expanding the expression (D1.4) in Discussion 1.1 in powers of v/c, show that at velocities low compared with the velocity of light the kinetic energy of a particle approaches the classical value, $\frac{1}{2}mv^2$.

1.2 It is tempting (but unwise) to assign to the wave associated with a material particle a 'frequency' given by the velocity of the particle divided by its de Broglie wavelength. Express the kinetic energy of a freely moving particle in terms of the frequency so defined, and compare the expression with that for a photon.

1.3 Suggest how two possible causes of the 'red shift' of light from distant stars might be distinguished by observations of the light:
(a) the Doppler effect on the light emitted by stars in an expanding universe;
(b) the cumulative effect of Compton scattering of the light by free electrons in outer space.

1.4 Using the classical expressions of Discussion 1.2 for the momentum and kinetic energy of the electron, and the expressions $p = h/\lambda$ and $E = h\nu$ for the momentum and kinetic energy of the photon, and recalling that $\lambda\nu = c$, derive the expression in Fig. 1.2 for the change of wavelength of a photon in Compton scattering.

2. Probability and Stationary States

IF the waves associated with a material particle are not the familiar electromagnetic waves, what are they? In an accepted interpretation of them, proposed by Max Born in 1926, they are *probability amplitude waves*. This interpretation has an analogue in one property of light waves, at any rate, and thus preserves some resemblance between the wave properties of photons and the wave properties of material particles.

Recall that the total amount of light passing per unit time through unit area of a plane is called the *intensity* of the light, and the intensity is proportional to the square of the amplitude of the light wave. In a particle picture of light, the amount of light must be proportional to the number of photons. In order to carry this interpretation over to a stream of material particles that exhibit wave-like phenomena, it is appropriate that the intensity of the wave, again proportional to the square of its amplitude, should measure the number of particles.

The interpretation of the wave as a probability amplitude wave follows at once. The probability of finding a particle in a chosen neighbourhood is proportional to the density of particles in that neighbourhood, and thus to the intensity of the wave in that neighbourhood. Like all others, this wave can be described mathematically as a function of space coordinates and the time. The first step toward exhibiting that function for any particular case is to construct Schrödinger's equation for the case, in the way that Chapter 4 will describe. Schrödinger's equation is a linear differential equation of the second order, whose solutions are the *wave functions* for the case. The wave function for a single particle is a function of the time and space coordinates for the particle, x, y, z, and t. And the square of the wave function associated with the particle measures, for each value of x and t, the probability that the particle is in the region of space between x and $x+\mathrm{d}x$ during the interval of time t to $t+\mathrm{d}t$.

When more than one particle must be considered, analogous procedures are available. Schrödinger's equation provides means for finding a wave function for N particles, which is a function of the time and of $3N$ space coordinates—enough for all the particles taken separately. The square of that wave function then measures the probability that particle No. 1 is in such-and-such a region, particle No. 2 is in some other specified region, and so on, all in the same little interval of time.

Return for the present to the simplest case, that of a single particle and its wave function. The central assumption of wave mechanics is that the wave function (along with a few procedures for using it) describes all that one can find out about the particle. Thus all that can be known about the position of the particle is the probability that it can be found in a certain region. The form of the wave function may imply that this probability is so great for such-and-such a region that it is nearly a certainty. On the other hand, the wave function may spread out in space as time goes on, or it may be diffracted into several beams, depending on the particular physical situation in question.

If it spreads out with time, the interpretation must be that there is a progressively larger region of space in which there is some probability of finding the particle. If it is diffracted into several beams, the particle may have taken several different courses, and one can determine only the relative

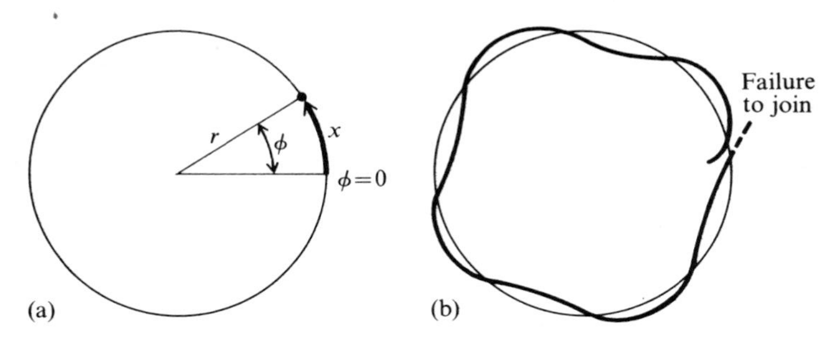

FIG. 2.1. The wave function for a particle constrained to move in a circle must be single-valued, and thus can have only certain wavelengths.

probability that it took one course as compared with another. In statistical language, the square of the wave function is a *distribution function* for the particle in space and time. In words that are more suggestive of the optical analogy, the wave function is a *probability amplitude* function and its square is a *probability intensity* function.

Among the most important wave functions are those analogous to two familiar sorts of mechanical and electrical waves: the running waves which do not change their form with time, and the standing waves produced by the superposition of two or more running waves. The vibration of a uniform string held at both ends in uniform tension is the most familiar mechanical example of a standing wave. Such mechanical or electrical waves can always be described by some function of the space coordinates alone, multiplied by $e^{i\omega t}$, and the analogous functions of wave mechanics are analogous in this respect also. Quantum-mechanical wave functions of this sort characterize any particle whose *total energy* (kinetic plus potential) is not changing with time.

As an example of such a case, examine the particle shown in Fig. 2.1, constrained to move in a circle. It seems a rather artificial example,† but it

† It is not quite as artificial as it seems, for it is the genesis of a way of treating the behaviour of electrons in metals by wave mechanics.

exhibits some of the properties of wave functions in a simple way. Think of the particle as tied to a central axle by a weightless rod, or as confined inside a tube of very narrow bore, bent into a circle.

Do not confuse this case with that of the hydrogen atom, with one electron revolving in an orbit around the nucleus. In the atom the electron is not constrained to move around the nucleus in a circle, nor indeed in an ellipse. Classical mechanics says that the electron will move in a circle or an ellipse (if the fact that it will radiate its energy is ignored), but not that it is constrained from outside to do so. Wave mechanics does not even say that the electron in an atom will do so. In the present problem, by contrast, the particle is rigidly restricted to move in a circle.

The solutions to Schrödinger's equation for this problem are the wave functions

$$\Psi = A \exp 2\pi \mathrm{i}(kx - \nu t), \tag{2.1}$$

where x is the distance around the circle from some fixed point on the circle, and ν, k, and A are constants. In order to see what such a function looks like, hold t fast,† choose some trial value of A and of k, substitute $x = r\phi$ (Fig. 2.1(a)), and plot either the real or the imaginary part of Ψ as a function of ϕ. For some value of k, that procedure will yield Fig. 2.1(b). In the latter figure the wave function (not the path of particle motion) is plotted as a wavy line about the circle used as a base line for $\Psi = 0$.

This looks like a wave except that it does not join onto itself after one period of 2π in ϕ. It can be made to join onto itself only by choosing the *wave number* (the number of waves per metre)

$$k = \pm\frac{n}{2\pi r}, \tag{2.2}$$

where n is an integer. In other words, these are the values of k which make Ψ single-valued in ϕ at constant t. Only this kind of wave function is suitable: Ψ will then be continuous, and Ψ^2 will be single-valued and thus interpretable as an unambiguous probability. In this way the problem yields a set of possible wave functions, which can be numbered with the positive or negative integer n:

$$\Psi_{\pm n} = A_{\pm n} \exp \mathrm{i}(\pm n\phi - 2\pi\nu t). \tag{2.3}$$

Notice that you can visualize other possible wave functions flowing from eqn (2.1) which are single-valued and continuous and which look like cos $n\phi$ over part of the circle and like sin $n\phi$ over the remainder. But the derivatives of such a function would exhibit discontinuities where its parts join. Schrödinger's equation permits such discontinuities only at points where the particle is subject to some discontinuous physical influence. If the potential energy of a particle differs in two adjacent regions, for example, jumping

† If a state of the particle is 'stationary', its description can be independent of time, as Chapter 4 will describe.

suddenly where the regions meet, one or more derivatives of the wave function may be discontinuous where the regions meet. In the present example the particle sees no such discontinuity within its one-dimensional world, the circle. But its world is restricted: the circumference of the circle is not infinitely long. It is that restriction which leads to the discreteness of its permitted states.

The wave functions can be connected with the behaviour of the particle by using the de Broglie relation between the wavelength of a wave function and the momentum of the particle,

$$p = h/\lambda. \tag{2.4}$$

Fig. 2.1(b) shows that the wavelength of the function is $\lambda = 2\pi r/\pm n$, and hence the momentum and the energy of the particle when it is described by the wave function $\Psi_{\pm n}$ are

$$p_{\pm n} = \frac{\pm nh}{2\pi r}; \tag{2.5}$$

$$E_{\pm n} = \frac{p^2}{2m} = \frac{n^2h^2}{8\pi^2r^2m}. \tag{2.6}$$

Thus to each of the possible wave functions there corresponds a value of the energy, and the possible values of energy form a discrete set, numbered by the integer n. For each possible value of the energy there are two wave functions in this case (except when $n = 0$), corresponding to the two directions of motion of the particle around the circle. When there are several possible states of a system with the same energy, each group of states with the same energy is called *degenerate*. The present case is one of *twofold degeneracy*.

Notice that the existence of discrete energy levels, and of a discrete set of possible states described by the wave functions, has come from the requirement that each wave function be single-valued and continuous. That requirement came in turn from the requirement that the square of the wave function should be interpretable without ambiguity as a probability. It is this last requirement that lies at the heart of the matter. The requirement can also be used to determine the constants $A_{\pm n}$ in the expression (2.3) for the wave functions, in the following way.

Since the wave functions (2.3) are complex—of the form $X+\mathrm{i}Y$, where X and Y are real functions of x and t—their squares will also be complex, and a complex number is not interpretable as a probability. The constant A cannot be chosen in such a way as to make the square of the wave function real for all values of x and t. In order to make probabilities real when wave functions are complex, wave mechanics interprets *the product of a wave function by its complex conjugate* as the probability distribution function:

$$\Psi\Psi^* \equiv (X+\mathrm{i}Y)(X-\mathrm{i}Y) = X^2+Y^2. \tag{2.7}$$

The probability distribution function obtained in this way from any of the wave functions (2.3) will be simply AA^*. Since this function is independent of ϕ, it asserts that the particle is equally likely to be found anywhere on the circle.

Now the probability that a particle will be somewhere in the space accessible to it is unity. In the case pictured in Fig. 2.1(a) the space accessible to the particle is restricted to a circular line. Thus each wave function must satisfy the requirement

$$\int_0^{2\pi} \Psi\Psi^* r \, d\phi = \int_0^{2\pi} AA^* r \, d\phi \qquad (2.8)$$
$$= 2\pi r AA^* = 1,$$

whence $AA^* = 1/2\pi r$. The constant A can be taken as the *real* constant $1/\sqrt{(2\pi r)}$. The choice of a complex value of A will merely change the phase of the wave. Choosing A by eqn (2.8), so as to give unit probability for finding the particle in its accessible space, is called *normalizing* the wave function.

When the space accessible to a particle is not restricted as it is in this case, so that the particle can be anywhere, the integration analogous to that in eqn (2.8) must be carried out over the whole of space. The integration limits are then infinite, and clearly most functions integrated between infinite limits cannot give finite answers. Again, therefore, the requirement that $\Psi\Psi^*$ shall be unambiguously interpretable as a probability distribution function picks out the wave functions that are acceptable as physically meaningful wave functions.

Here two common physical situations must be distinguished. If a particle is attracted to some region of space, and has too little energy to escape permanently from that region, it may stray but it will always return. It is more likely to be found in that region than elsewhere, and the probability of finding it at a great distance will approach zero as the distance from that region increases. An electron bound to an atom is an important example. In such a case the integration analogous to eqn (2.8) over all space must yield a finite answer. Then the acceptable solutions to Schrödinger's equation are *quadratically integrable* functions, and they will usually form a discrete set, to each member of which a definite value of the energy corresponds.

If a particle acquires enough energy to escape permanently from confinement, it will still be found more probably in the attractive region, but the probability that it will be found at an infinite distance no longer vanishes. Thus when an electron is ejected from an atom, the appropriate *continuum functions* will not be quadratically integrable and will not form a discrete set.

PROBLEMS

2.1 What is the phase velocity of the wave described by the wave function (2.3)?

2.2 Discuss the limitations, if any, to the quadratic integrability over the range $x = 0$ to ∞ of functions of the type $e^{-\alpha x}(a_0 + a_1 x + \ldots + a_n x^n)$, where n is a finite integer.

3. Uncertainty and the Particle in a Box

A PRINCIPLE that lies deep in the philosophical groundwork of quantum mechanics is the *uncertainty principle*, proposed by Werner Heisenberg in 1927. It examines the combined accuracy with which measurements of two different dynamical properties of a system can be made. For example, it says that the position of a particle and the momentum of the particle cannot both be determined exactly. And it assigns a relation between the uncertainties which must inhere in any attempt to measure both of these quantities. The principle has far-reaching consequences in the observed behaviour of matter; it is of greater practical importance than may appear at first.

Classical mechanics offers no objection to such joint measurements; both position and momentum can in principle be determined to any desired degree of accuracy. Quantum mechanics asserts, in this instance, that the uncertainty Δx in a measurement of the position x of the particle and the uncertainty Δp_x in a measurement of its component of momentum p_x along the x direction must obey the relation

$$\Delta x \, \Delta p_x \sim h. \tag{3.1}$$

The product of the two uncertainties can never be made less than a quantity of the order of Planck's constant.†

Often the relation (3.1) is made reasonable by analysing imaginary idealized experiments. A feature of many of these *thought experiments* is that the attempt to measure one of the quantities changes the value of the other: the measured system cannot be contemplated independently of the measuring system.

Instead of analysing such experiments, be content here to examine the meaning of the uncertainty principle in the case of a particular dynamical system, that of *the particle in a box*. It is simplest to choose a one-dimensional box: the particle is constrained to move along a line of length l. In other words, the thin tube used to picture circular motion in the last chapter is cut at one point, bent straight, and capped at both ends. The particle can only shuttle back and forth with constant kinetic energy along the line, reversing the direction of its momentum every time it comes to an end of the line.

The wave functions for this case are the standing-wave counterparts of the running waves that describe the particle on a circle:

$$\Psi_n = A_n \sin \frac{n\pi x}{l} \exp -2\pi i \nu t$$

† Different statements of the uncertainty principle flow from different definitions of 'the uncertainty'. The statistical definition described in the appendix to Chapter 8 leads to the statement $\Delta x \, \Delta p_x \geq h/2\pi$.

for x in the range 0 to l,

$$\Psi_n = 0 \tag{3.2}$$

for x outside the range 0 to l. The choice of these wave functions is dictated by the requirements that there can be no probability of finding the particle beyond the ends of the line, and that the function must still be continuous at the ends. But at those two points the derivatives of the function need not be continuous, because the particle meets abrupt obstructions there.

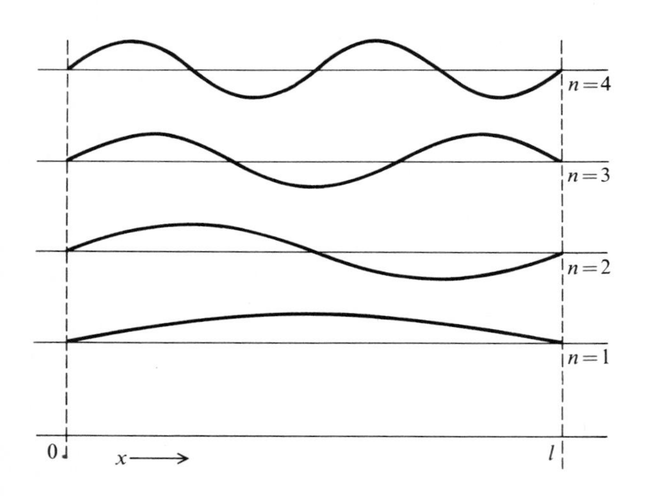

FIG. 3.1. The first few wave functions for the particle in a one-dimensional box.

For the first few values of n the wave functions are drawn in Fig. 3.1. The reasoning used in Chapter 2 shows that the wavelengths are $\lambda = 2l/n$, that the momenta are therefore $nh/2l$, and thus that the energy levels are $n^2h^2/8ml^2$. If the potential energy of a particle inside the box is taken to be zero, the energies given by this expression are all kinetic.

Notice particularly the fact, characteristic of wave mechanics, that the energy corresponding to the wave function does not depend on the value of A_n. In this respect the waves are quite different from those of ordinary mechanics and of alternating-current circuitry. Here the energy does not depend on the amplitude. The appropriate value of A_n is that which normalizes the wave function, making $\Psi\Psi^*$ interpretable as a probability, as Chapter 2 described.

Some other features of wave mechanics, which are very generally applicable, appear clearly in this example. Notice that the value $n = 0$ is useless because the corresponding wave function (3.2) would vanish everywhere; and the probability interpretation would be that the particle is nowhere—that there is no particle. Then the first admissible wave function—that with the lowest energy—is that in which $n = 1$, with a corresponding energy $h^2/8ml^2$. Thus the particle cannot have zero energy and cannot be at rest.

Clearly this feature of the example comes from the fact that the particle is confined to the length l. The formula for energy levels shows that as l gets bigger this lowest admissible energy, and the differences between all permitted energies, get smaller. Conversely, the smaller the box the higher the lowest permitted kinetic energy of the particle and the greater the separation of permitted energies.

A particle confined to move in one dimension between two reflecting walls is a very artificial construction, of course, but some of its properties suggest those of particles found in real situations. An atom bound within a molecule, for example, has a potential energy that depends on its position with respect to its fellows in the molecule. The atom will seek those regions where its potential energy is lowest, and the probability that it will be found in those regions will tend to be high. Because of this localization, which is analogous to the stricter localization of a particle in a box, wave mechanics will usually assign a lowest possible kinetic energy to the atom, even when the molecule as a whole is at rest, which corresponds to a wave function called the *ground state* of the atom. Again this energy will be larger the more narrowly the potential energy of the atom tries to restrict its position.

For the particle in a box the same qualitative result follows from the uncertainty principle. If the particle is certainly in the box, then the uncertainty in its position is $\Delta x = l$, a finite quantity. The principle asserts that the product of that uncertainty and the uncertainty in the momentum of the particle is finite, and thus the uncertainty in momentum must also be finite. If the momentum were certainly zero, there would be no uncertainty in it. The particle must therefore have some mean absolute momentum, and with it a mean kinetic energy.

This reasoning can be pushed one step further by examining directly the uncertainty in the momentum of the particle. Since the particle may at any time have either a momentum $+p$ or $-p$, depending on which way it is moving, the uncertainty in p_x is $2p$. By the de Broglie relation $p = h/\lambda$. Since the permitted wavelengths are $\lambda = 2l/n$, $\Delta p_x = nh/l$, where the integer n is at least unity. Hence $\Delta x \, \Delta p_x = nh$; in other words $\Delta x \, \Delta p_x$ cannot be less than h.

The uncertainty principle not only connects uncertainties in position and momentum but also connects uncertainties in energy and time. By deriving the latter relation $\Delta E \, \Delta t \sim h$, from the former relation $\Delta p \, \Delta x \sim h$, in a particular example, Discussion 3.1 may help to make the two relations equally credible.

But the relation $\Delta E \, \Delta t \sim h$ is more general than this derivation implies. The energy in question is not restricted to kinetic energy. The relation connects

Discussion 3.1

UNCERTAINTIES IN ENERGY AND TIME

Consider the problem of finding out when a particle with associated uncertainties Δx and Δp_x will arrive at a certain place. Fig. 3.2 shows that at some

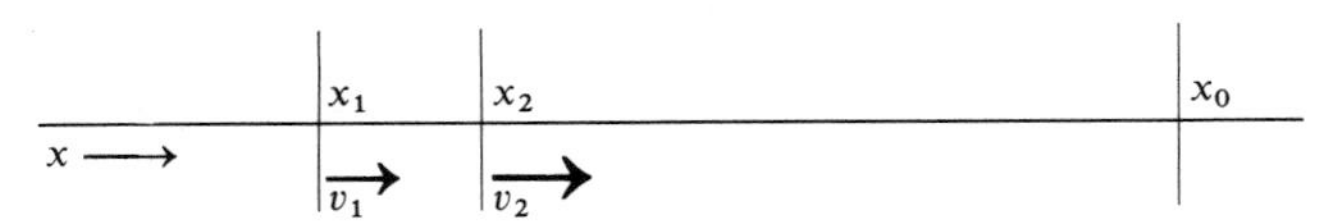

FIG. 3.2. The connection between uncertainties in the position and momentum of a particle implies a connection between uncertainties in its kinetic energy and its time of arrival at x_0.

time $t = 0$ the uncertainty of position will allow the particle to be located anywhere between x_1 and x_2 (where $\Delta x = x_2 - x_1$). In consequence of the associated uncertainty in its momentum, the velocity of the particle may have any value between v_1 and v_2 (where $\Delta p_x/m = v_2 - v_1$). The uncertainty in its time of arrival at x_0 will be the difference between the longest possible time, $(x_0 - x_1)/v_1$, and the shortest possible time, $(x_0 - x_2)/v_2$. Hence

$$\Delta t = \frac{x_0 - x_1}{v_1} - \frac{x_0 - x_2}{v_2},$$

and if $v_2 - v_1$ is small compared with their mean value $v = \frac{1}{2}(v_1 + v_2)$, then

$$\Delta t = \frac{x_2 - x_1}{v} = \frac{\Delta x}{v}.$$

But the uncertainty in momentum implies an uncertainty in kinetic energy

$$\Delta E = \tfrac{1}{2}m(v_2^2 - v_1^2)$$
$$= \frac{v_2 + v_1}{2} m(v_2 - v_1) = v\,\Delta p_x,$$

and thus

$$\Delta E\,\Delta t = \Delta x\,\Delta p_x \sim h.$$

any uncertainty in the value of the total energy of some system or part of a system, and any uncertainty in the time at which that energy is measured.

For example, certain observed spectral lines arise from atomic or molecular transitions from states of very short life. If the life of such a state is short enough, the uncertainty Δt in the time at which it makes its transition to another state, absorbing or emitting a photon, is necessarily small. Thus the spectral line is broadened by operation of the principle $\Delta E\,\Delta t \sim h$, and the relation $\Delta E = h\,\Delta \nu$ for the emitted photons. The broadening is noticeable experimentally when the initial states have lifetimes shorter than 10^{-10} s.

PROBLEMS

3.1 Normalize the wave functions for the particle in a one-dimensional box.

3.2 Show how the wave functions for the particle in a one-dimensional box can be constructed by considering them to be standing waves formed by superposing two of the running wave functions obtained in Chapter 2 for the particle on a circle.

4. Schrödinger's Equation

SCHRÖDINGER's equation furnishes the means for constructing the wave functions for mechanical systems. There are many ways of making it seem reasonable—some of them statistical, others developing an analogy between the mathematical formulations of classical mechanics and optics. All provide a helpful insight into its meaning.

Alternatively, Schrödinger's equation can be taken as a postulate from which to develop a form of mechanics whose reasonableness consists in its implying consequences that agree with experiment. This chapter will adopt the latter viewpoint, since the former would require a preliminary excursion through classical mechanics, statistical mechanics, and optics.

It may help you to accept Schrödinger's equation in a friendly spirit by seeing how it fits together with some properties of the wave functions for the particle on a circle, discussed in Chapter 2,

$$\Psi = A \exp 2\pi i(kx - \nu t), \tag{4.1}$$

where distance along the circle, $r\phi$, is now denoted by x. Recall (Chapter 2, eqns 2.2 and 2.5) that the momentum is related to the wave number k by

$$p_x = hk, \tag{4.2}$$

and notice that in consequence the momentum can be found by differentiating the wave function with respect to x,

$$\frac{\partial \Psi}{\partial x} = 2\pi i A k \exp 2\pi i(kx - \nu t). \tag{4.3}$$

This permits you to write

$$\frac{h}{2\pi i}\frac{\partial}{\partial x}\Psi = p_x \Psi, \tag{4.4}$$

and to read that statement in the words, 'The performance of the operation $(h/2\pi i)(\partial/\partial x)$ on the wave function Ψ is equivalent to multiplying the wave function by p_x'.

Look now at the constant ν. The analogy between photons and material particles, used in developing quantum-mechanical ideas in Chapter 1, suggests setting the energy

$$E = h\nu. \tag{4.5}$$

Then the energy of the particle can be obtained from the wave function by differentiating it with respect to t, much as the momentum was obtained by differentiation with respect to x:

$$-\frac{h}{2\pi i}\frac{\partial}{\partial t}\Psi = E\Psi. \tag{4.6}$$

It is important to notice (Problem 2.2) that this 'frequency' ν is not equal to the velocity of the particle divided by its de Broglie wavelength, except when the particle is a photon.

Consider now the energy equation that this particle would obey according to classical mechanics:

$$E = \tfrac{1}{2}mv^2 = \frac{p_x^2}{2m}. \tag{4.7}$$

When both sides of it are multiplied by Ψ, the quantities E and p_x can be replaced by operations—*wave-mechanical operators.* The equation then becomes

$$\begin{aligned}\frac{-h}{2\pi i}\frac{\partial\Psi}{\partial t} &= \frac{1}{2m}\left(\frac{h}{2\pi i}\frac{\partial}{\partial x}\right)\left(\frac{h}{2\pi i}\frac{\partial}{\partial x}\right)\Psi \\ &= \frac{-h^2}{8\pi^2 m}\frac{\partial^2\Psi}{\partial x^2}.\end{aligned} \tag{4.8}$$

Eqn (4.8) is Schrödinger's equation for a particle moving freely but in only one dimension along the coordinate x. It is an energy equation, obtained by replacing the quantities p_x and E in the classical energy equation by differential operators. Solutions of this differential equation that obey physically reasonable conditions are the wave functions Ψ for the particle. The physically reasonable conditions are those which permit the function $\Psi\Psi^*$ to be interpreted as a probability, as Chapter 2 has described.

The preceding paragraphs have started with wave functions and arrived at Schrödinger's equation. Now examine the reverse procedure: arriving at the wave functions by starting with Schrödinger's equation. This is what must in fact be done in a physical problem; it was done to derive the functions (4.1) in the first place.

Since eqn (4.8) contains derivatives of a function with respect to x and t, its solutions Ψ must be functions of both these variables. It is a common device, in solving linear partial differential equations, to seek solutions that are sums or products of functions of each of the variables taken separately—the method of 'separation of variables'. In this case assume that it will be possible to find solutions of the form

$$\Psi(x, t) = \psi(x)f(t). \tag{4.9}$$

Substituting (4.9) into (4.8), and dividing both sides by Ψ, yields

$$\frac{-h}{2\pi i f}\frac{\mathrm{d}f}{\mathrm{d}t} = \frac{-h^2}{8\pi^2 m\psi}\frac{\mathrm{d}^2\psi}{\mathrm{d}x^2}. \tag{4.10}$$

Since the left side is a function of t only, and the right side is a function of x only, each side must be equal to a constant K, independent of both x and t.

This observation provides two ordinary linear differential equations, one for f and one for ψ. The equation for f,

$$\frac{-h}{2\pi \mathrm{i} f}\frac{\mathrm{d}f}{\mathrm{d}t} = K, \tag{4.11}$$

can be solved by simple quadrature:

$$\frac{\mathrm{d}f}{f} = -\frac{2\pi\mathrm{i}}{h}K\,\mathrm{d}t, \quad \text{or}$$

$$f = (\text{constant})\exp -2\pi\mathrm{i}Kt/h. \tag{4.12}$$

The equation for ψ is a linear differential equation of the second order with constant coefficients,

$$\frac{\mathrm{d}^2\psi}{\mathrm{d}x^2}+\frac{8\pi^2 mK}{h^2}\psi = 0. \tag{4.13}$$

The solutions of such equations always have the form $\psi = A\mathrm{e}^{\alpha x}$, where A and α are constants. Substituting this form into (4.13), and dividing by ψ, shows that it is a solution, for any value of A, when

$$\alpha^2+\frac{8\pi^2 mK}{h^2} = 0, \quad \text{or} \quad \alpha = \frac{\pm 2\pi\mathrm{i}}{h}\sqrt{(2mK)}. \tag{4.14}$$

Here are two distinct solutions of (4.13), one for each choice of the sign of α in (4.14), and each solution can be multiplied by any constant A. Clearly the sum of two solutions is also a solution, and thus the most general solution for (4.13) is

$$\psi = A\exp\frac{2\pi\mathrm{i}}{h}\sqrt{(2mK)}x+B\exp\frac{-2\pi\mathrm{i}}{h}\sqrt{(2mK)}x. \tag{4.15}$$

Recalling the original separation of variables in (4.9), you can now combine (4.12) and (4.15) into the solutions of (4.8) that you sought:

$$\Psi = A\exp\frac{2\pi\mathrm{i}}{h}\{\sqrt{(2mK)}x-Kt\}+$$

$$+B\exp\frac{2\pi\mathrm{i}}{h}\{-\sqrt{(2mK)}x-Kt\}. \tag{4.16}$$

Applying to this wave function Ψ the momentum operator (4.4), first with $B = 0$ and then with $A = 0$, provides $(1/\Psi)(h/2\pi\mathrm{i})(\partial/\partial x)\Psi = \sqrt{(2mK)}$ and $-\sqrt{(2mK)}$. If K is identified with the energy E, the relation (4.7) shows that the operator has indeed yielded the momentum; the two signs correspond with the two possible directions of motion of the particle.

Do not make the mistake of supposing that anything substantial has been proved by this chase from wave function to Schrödinger's equation and back to wave function. The chase has merely shown how everything in the last few chapters hangs together, and has exemplified the mathematical technique for solving Schrödinger's equation. In the context of this monograph, that equation must be taken as one of the postulates of the system of mechanics under construction. This preamble may help you to accept the procedure for writing Schrödinger's equation in more general cases.

The first step in setting up Schrödinger's equation for a system is to write its total energy in terms of the time and of the momenta and coordinates of its parts. The second step is to substitute, for the energy and momenta in that energy equation, the quantum mechanical operators. These two steps correspond to steps (4.7) and (4.8) in the case already discussed. In that case the energy of the particle was entirely kinetic, and there was no need to add a term to represent the potential energy.

When a potential energy term must be added, what is the corresponding operation on the wave function? It is simply multiplication of the wave function by the potential energy as a function of the coordinates. 'Operation on the wave function by x' means 'multiplication by x'. A similar statement applies to any function of the coordinates: operation on the wave function by any function of the coordinates is the same as multiplying the wave function by that function of the coordinates.

In the more general case of a system of particles that interact with one another, and also perhaps with externally applied forces, the total energy of the system at any instant is the sum of its kinetic energy and its potential energy. The kinetic energy will be

$$\begin{aligned} U_{\text{kin}} &= \tfrac{1}{2}m_1 v_1^2 + \tfrac{1}{2}m_2 v_2^2 + \dots \\ &= \frac{p_1^2}{2m_1} + \frac{p_2^2}{2m_2} + \dots, \end{aligned} \tag{4.17}$$

summed for all the particles. The potential energy of the system, in all the cases of interest in this series of monographs, will not depend on the momenta of its parts but only on the coordinates of the particles (and the values of any externally applied forces):

$$U_{\text{pot}} = U_{\text{pot}}(x_1, y_1, z_1; x_2, y_2, z_2; \dots). \tag{4.18}$$

The classical energy equation is then

$$U_{\text{kin}} + U_{\text{pot}} = E, \tag{4.19}$$

where $U_{\text{kin}} + U_{\text{pot}}$ is written out in terms of the momenta and coordinates of the system, and when so written out is called the *Hamiltonian* of the system, and is denoted by H. With this notation eqn (4.19) can be written in the simple form

$$H = E. \tag{4.20}$$

Now multiply both sides of the equation by Ψ and substitute the quantum mechanical operators. Since $p_1^2 = p_{x1}^2+p_{y1}^2+p_{z1}^2$, etc., the term U_{kin}, the kinetic energy, requires the substitutions

$$p_{x1} \to \frac{h}{2\pi i}\frac{\partial}{\partial x_1}, \qquad p_{y1} \to \frac{h}{2\pi i}\frac{\partial}{\partial y_1},$$
$$p_{z1} \to \frac{h}{2\pi i}\frac{\partial}{\partial z_1}, \text{ etc.} \tag{4.21}$$

The operators thus obtained are

$$\frac{p_1^2}{2m_1} \to \frac{-h^2}{8\pi^2 m_1}\left(\frac{\partial^2}{\partial x_1^2}+\frac{\partial^2}{\partial y_1^2}+\frac{\partial^2}{\partial z_1^2}\right), \text{ etc.} \tag{4.22}$$

The part of the operator (4.22) in parentheses is called the 'Laplacian' operator and is often written for convenience

$$\nabla_1^2 = \frac{\partial^2}{\partial x_1^2}+\frac{\partial^2}{\partial y_1^2}+\frac{\partial^2}{\partial z_1^2}. \tag{4.23}$$

After making the further substitution of the operator for the energy

$$E \to \frac{-h}{2\pi i}\frac{\partial}{\partial t}, \tag{4.24}$$

as in eqn (4.6) for the earlier example, the energy equation reads

$$\frac{-h^2}{8\pi^2}\left[\frac{\nabla_1^2}{m_1}+\frac{\nabla_2^2}{m_2}+\dots\right]\Psi + U_{\text{pot}}\,.\,\Psi = \frac{-h}{2\pi i}\frac{\partial}{\partial t}\Psi. \tag{4.25}$$

Eqn (4.25) can be abbreviated to read

$$H\Psi = E\Psi, \tag{4.26}$$

where H and E denote the Hamiltonian operator and the energy operator, respectively. It is the wave-mechanical analogue of eqn (4.20).

Written for N particles, the resulting equation is a linear partial differential equation in $3N+1$ variables: the three space coordinates of each particle and the time. Solving it exactly is usually impossible, and even in the soluble cases invokes mathematical techniques that are not elementary. In order to make progress despite these difficulties, the soluble cases are used to give insight into the insoluble cases and to provide means for making approximate calculations. But often the methods of approximation also invoke non-elementary mathematical tools.

The resulting mathematical character of wave mechanics may seem repellent in a physical theory, but there is no help for it. Indeed, that mathematical character lies deeper than mere technique. Wave-particle duality (Chapter 1) prevents one from using familiar ideas of how a particle behaves, and leaves one with the picture of a wave function whose properties are mathematical properties. In order to gain notions of how wave functions behave, you must examine the mathematical properties of quite a few of them. Then you can achieve some of the same intuitive feeling for quantum mechanical behaviour that you already have for classical mechanical behaviour.

One important simplification can be made at once in eqn (4.26). In the systems discussed in this series of monographs, the Hamiltonian is a function of the coordinates and momenta but not of the time. Hence eqn (4.26) can be broken in two by the method of separation of variables, used in eqn (4.9) of the earlier case,

$$\Psi = \psi(x_1 \ldots)f(t). \tag{4.27}$$

Just as before, $H\Psi = (H\psi)f$, and $E\Psi = (-h/2\pi\mathrm{i})(\mathrm{d}f/\mathrm{d}t)\psi$. Substituting those expressions into eqn (4.26), and dividing both sides by Ψ, provides

$$\frac{H\psi}{\psi} = \frac{-h}{2\pi\mathrm{i}}\frac{1}{f}\frac{\mathrm{d}f}{\mathrm{d}t}, \tag{4.28}$$

and since the left side is a function of the coordinates only, and the right side is a function of the time only, each side equals a constant K. The equation for f is then the familiar eqn (4.11), whose solution is

$$f = (\text{constant})\exp -2\pi\mathrm{i}Kt/h. \tag{4.29}$$

Again K can be identified with the total energy of the particle, as the next chapter will show. Thus the other half of eqn (4.28), reading $H\psi = K\psi$, becomes

$$H\psi = E\psi, \tag{4.30}$$

where E no longer denotes an 'energy operator' but rather the energy itself. And in consequence of eqn (4.27), the solutions of eqn (4.26) can be written

$$\Psi = \psi \exp -\frac{2\pi\mathrm{i}}{h}Et. \tag{4.31}$$

Eqn (4.30) looks like eqn (4.26), but notice carefully two differences. In the first place, the function ψ is not time-dependent; it is a function only of the coordinates. In the second place, E is no longer a differential operator; it is a constant. Eqn (4.30) is called the 'time-independent Schrödinger equation'. Solutions of it that permit a probability interpretation are the time-independent wave functions for the system, describing the states that the system can adopt. Usually such solutions are possible only for special values of the constant E. Those values are the possible energy levels of the system, and this scheme of solution will associate certain states of the system with certain values of the energy.

PROBLEMS

4.1 What feature must be exhibited by any wave function, for a particle moving in one dimension, at any point of space at which the potential energy of the particle is equal to its total energy?

4.2 If you are acquainted with the meaning of the operator ∇ in vector analysis, answer the question, 'What is the spatial symmetry of the part of the Schrödinger equation that represents the kinetic energy'?

5. Wave Functions, Operators and the Harmonic Oscillator

THE most important properties of wave functions are associated with the interpretation of their squares, $\Psi\Psi^*$, as measuring probabilities in the way described in Chapter 2. That interpretation makes it possible to calculate what the observable properties of a system are. But you must recognize that, according to the uncertainty principle described in Chapter 3, the observable properties may be somewhat indefinite.

'Where is the particle'? You can say two things that provide a partial answer to that question. In the first place, you can find the relative probability that it is one place or another if you know its *state*—its wave function—as Chapter 2 has already described. Hence, in particular, you can find its *most probable location.*

In the second place, you can find its *average* location by writing down each possible location, weighting that location by the probability that the particle is in it, and adding all the results. Thus, for example, the average x coordinate of a single particle, restricted to move in a line and known to be in a state described by the wave function Ψ, is

$$\bar{x} = \frac{\int x\Psi^*\Psi \, dx}{\int \Psi^*\Psi \, dx}, \tag{5.1}$$

where the integration is carried over all values of x accessible to the particle. If the wave function has been normalized (Chapter 2), then $\int \Psi^*\Psi \, dx = 1$, and eqn (5.1) can be written

$$\bar{x} = \int x\Psi^*\Psi \, dx. \tag{5.2}$$

If the particle is only one member of a system of many particles, free to move in all three dimensions, Ψ is a function of the coordinates of all the particles, and

$$\bar{x}_1 = \int x_1\Psi^*\Psi \, d\tau, \tag{5.3}$$

where the integration is carried over all the coordinates of all the particles. It is a multiple integration; $d\tau$ stands for $dx_1 \, dy_1 \, dz_1 \, dx_2 \, dy_2 \, dz_2 \ldots$.

Now consider the question, 'What is the momentum of the particle?' The probability interpretation applies to its location, but says nothing directly about its momentum. It is natural to guess that at least the *average* momentum might be calculated in some way analogous to the calculation of the average position. But Chapter 4 has made available only an operator, not a quantity, to represent the instantaneous momentum. It is tempting to write, by analogy with eqn (5.3), $\bar{p}_{x1} = \int p_{x1}\Psi^*\Psi \, d\tau$, and to use the quantum-mechanical operator $(h/2\pi i)(\partial/\partial x_1)$ for p_{x1} under the integral sign.

That guess is essentially right, but a new precaution is needed. In the expression used for obtaining $\bar{x}_1$, it made no difference in what order x_1, Ψ, and Ψ^* appeared under the integral sign because all these quantities are *commutative*: $x_1\Psi = \Psi \, . \, x_1$, for example. But when the operator does not commute with the object on which it operates, the order of writing makes a difference. For example, if the state under examination is the ground state of a particle in a box (Chapter 3), then

$$p_{x1}\Psi^*\Psi = (h/2\pi i)[(2A\pi/l)\sin(\pi x/l) \, . \, \cos(\pi x/l)],$$

$\Psi^* p_{x1}\Psi$ has half that value, and $\Psi^*\Psi p_{x1}$ has no meaning.

It is a postulate of wave mechanics that the correct form to use, with any operator G presenting such an ambiguity, is

$$\bar{G} = \int \Psi^* G \Psi \, d\tau. \tag{5.4}$$

Using this form for the mean momentum of the particle in a box, you find that it is zero. This result makes physical sense because the particle is moving in one direction as often as in the other. Using this form for the mean of the *square* of the momentum, you obtain a nonvanishing answer that can be checked against the value of the kinetic energy.

A mean value calculated by eqn (5.4) is called an *expectation value*. It is the value of an observable property that would be obtained by measuring the property in a great many systems that are identical, so far as they can be made identical, and then averaging those measurements.

Turn now to examine the expectation value of the *energy* of a system. The preceding discussion has exhibited two operators for the total energy (eqn 4.26): the operator H and the quantity E. In order to leave no discrepancy between them, the expectation value $\bar{H}$ of H must be made equal to the energy E. The identity of these two quantities is established in Discussion 5.1.

Examine as an example of these methods, useful in its own right, the case of the harmonic oscillator, with which your earlier studies of classical mechanics have already made you familiar. A particle oscillates back and forth along the x direction about the position $x = 0$, under a restoring force $F_x = -kx$. Its potential and kinetic energies are

$$U_{\text{pot}} = \tfrac{1}{2}kx^2, \qquad U_{\text{kin}} = \tfrac{1}{2}m\left(\frac{dx}{dt}\right)^2 = \frac{p_x^2}{2m}. \tag{5.5}$$

Discussion 5.1

THE ENERGY OF A WAVE-MECHANICAL SYSTEM

When a mechanical system is in a state whose wave function has the form $\Psi = \psi(x_1 \ldots) \,.\, f(t)$, the time-dependent part of that function is

$$f = (\text{constant}) \,.\, \exp(-2\pi \mathrm{i} Kt/h),$$

as Chapter 4 has already discussed. Applying the rule $\bar{H} = \int \Psi^* H \Psi \, \mathrm{d}\tau$, you obtain

$$\bar{H} = \int \psi^* \exp(2\pi \mathrm{i} Kt/h) H\psi \exp(-2\pi \mathrm{i} Kt/h) \, \mathrm{d}\tau.$$

Since H is independent of t, you can cancel one exponential against the other and write

$$\bar{H} = \int \psi^* H \psi \, \mathrm{d}\tau.$$

Since $H\psi = K\psi$ and K is a constant, you have $\bar{H} = K \int \psi^* \psi \, \mathrm{d}\tau$. Thus if ψ is normalized, $\bar{H} = K$. In other words, you can identify K with the energy E, and justify the way the time-dependent Schrödinger equation was written in eqn (4.30).

In classical mechanics its equation of motion follows from Newton's observation that the force on the particle equals its mass times its acceleration:

$$m \frac{\mathrm{d}^2 x}{\mathrm{d}t^2} = -kx. \tag{5.6}$$

The solutions to this equation give its position as a function of time,

$$x = A \cos 2\pi\nu(t - t_0), \tag{5.7}$$

where A is the amplitude of the oscillation and t_0 is its phase. The quantity ν, its frequency, is

$$\nu = \frac{1}{2\pi} \sqrt{\left(\frac{k}{m}\right)}, \tag{5.8}$$

and you can use this relation to write eqn (5.6) in terms of a single descriptive parameter:

$$\frac{\mathrm{d}^2 x}{\mathrm{d}t^2} + 4\pi^2 \nu^2 x = 0. \tag{5.9}$$

Using the solution (5.7) in (5.5), you find

$$\begin{aligned} U_{\text{pot}} &= \tfrac{1}{2} A^2 k \cos^2 2\pi\nu(t - t_0), \\ U_{\text{kin}} &= \tfrac{1}{2} A^2 k \sin^2 2\pi\nu(t - t_0). \end{aligned} \tag{5.10}$$

Thus the potential and kinetic energies oscillate out of phase but with the same amplitude, so that their average values are the same. Their sum, the total energy, is $\frac{1}{2}kA^2$.

In order to carry out the wave mechanical calculation, starting from (5.5), add U_{kin} and U_{pot}, substitute the momentum operator for p_x, and write the time-independent Schrödinger equation,

$$H\psi = \frac{-h^2}{8\pi^2 m}\frac{d^2\psi}{dx^2} + \tfrac{1}{2}kx^2\psi = E\psi. \tag{5.11}$$

This equation can be made to look simpler by lumping the constants together into two parameters,

$$\lambda \equiv 8\pi^2 mE/h^2, \quad \text{and} \quad \alpha^2 \equiv 4\pi^2 mk/h^2. \tag{5.12}$$

Notice that, by using (5.8), α can be written rationally in terms of the classical frequency, $\alpha \equiv 4\pi^2 m\nu/h$. Eqn (5.11) then reads

$$\frac{d^2\psi}{dx^2} + (\lambda - \alpha^2 x^2)\psi = 0. \tag{5.13}$$

Pause at this point to compare the wave equation (5.13) with eqn (5.9), which provided the solutions to the classical problem. Both are linear differential equations of the second order. In the classical equation a space coordinate depends on the time; in the quantum equation a wave function depends on a space coordinate. The classical equation is easy to solve; the quantum equation is not, because it is a differential equation with variable rather than constant coefficients. You seek the most general solution of the classical equation, assured that any solution has physical significance. You seek only special solutions of the quantum equation: those whose squares can be integrated over the range of x from $-\infty$ to $+\infty$ to yield a finite result. You have reason to believe that only certain values of the parameter λ will permit this; those values will determine the permitted energy levels through the relations (5.12).

The details of how to obtain the desired solutions to eqn (5.13) need not be repeated, but it may be worthwhile to examine hastily the general mathematical ideas of the procedure.† Notice that for very large values of x the coefficient of ψ in eqn (5.13) is dominated by $\alpha^2 x^2$, and that $A \exp -\alpha x^2/2$ is an asymptotic solution to the equation for large x, since

$$(d^2/dx^2)\exp -\alpha x^2/2 = (\alpha^2 x^2 - \alpha)\exp -\alpha x^2/2.$$

Indeed for the special value $\lambda = \alpha$, it is an exact solution. Moreover it is an acceptable solution, since it is quadratically integrable

$$\int_{-\infty}^{+\infty} A^2 \exp(-\alpha x^2/2)\, dx = A^2\surd(\pi/\alpha)$$

and the function therefore permits a probability interpretation.

† A closely analogous procedure is used in a case of greater practical importance, that of the 'hydrogen-like atom', and is described in detail in *The nature of atoms*, Chapter 7.

The remaining acceptable solutions are found by examining functions formed by multiplying this function by a power series in x:

$$\psi = A\exp(-\alpha x^2/2)(a_0+a_1x+a_2x^2+\ldots+a_nx^n+\ldots). \tag{5.14}$$

When this expression for ψ is used in eqn (5.13), you find in the first place that the power series must terminate—the bracketed part of the expression must be a polynomial—in order to secure quadratic integrability. You then find that there is one solution, corresponding to a different value of λ, for each value of n chosen to terminate the series. Those values of λ are in fact $\lambda_n = \alpha(2n-1)$, and hence by the relation (5.12) the energy levels are

$$E_n = (n+\tfrac{1}{2})h\nu. \tag{5.15}$$

The corresponding wave functions are of the form

$$\psi_n = A_n\exp(-\alpha x^2/2)(a_0+a_1x+\ldots+a_nx^n). \tag{5.16}$$

The bracketed polynomials, with values of the a's chosen to satisfy eqn (5.13), are called the 'Hermite polynomials'.

Since the functions are quadratically integrable, the coefficient A_n can be chosen so as to normalize them. The first four of these functions (with their normalizing coefficients) and the corresponding energies are

$$\psi_0 = \left(\frac{\alpha}{\pi}\right)^{\frac{1}{4}}\exp -\alpha x^2/2, \qquad E_0 = \tfrac{1}{2}h\nu,$$

$$\psi_1 = \left(\frac{4\alpha^3}{\pi}\right)^{\frac{1}{4}}x\exp -\alpha x^2/2, \qquad E_1 = \tfrac{3}{2}h\nu,$$

$$\psi_2 = \left(\frac{\alpha}{4\pi}\right)^{\frac{1}{4}}(1-2\alpha x^2)\exp -\alpha x^2/2, \qquad E_2 = \tfrac{5}{2}h\nu,$$

$$\psi_3 = \left(\frac{9\alpha^3}{\pi}\right)^{\frac{1}{4}}\left(x-\frac{2\alpha}{3}x^3\right)\exp -\alpha x^2/2, \qquad E_3 = \tfrac{7}{2}h\nu. \tag{5.17}$$

Fig. 5.1 shows (dashed lines) the first three of these functions plotted against x, using lines at the energy levels appropriate to them as base lines. Their squares, the probability functions, are also plotted (solid lines); and the parabolic potential energy curve $U_{\mathrm{pot}} = \frac{1}{2}kx^2$ is shown as well, to assist comparison with classical expectations.

Notice in the first place in Fig. 5.1 a general similarity of these wave functions to those for the particle in a box (Fig. 3.1). In both cases the functions oscillate about the horizontal axis, crossing it an increasing number of times as n, the *quantum number*, increases. In both cases there is a *zero-point* energy; the lowest permitted value of the kinetic energy is not zero.

In fact you can think of the two cases as similar if you think of them both

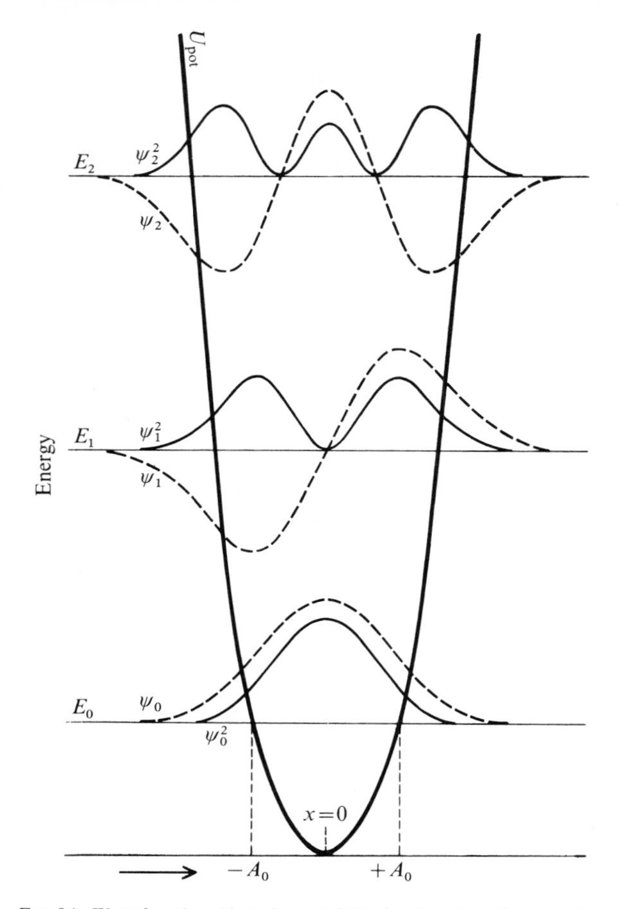

FIG. 5.1. Wave functions (dotted), probability functions (solid), classical potential energy (parabola), and energy levels (horizontal) for the harmonic oscillator.

as cases of particles in *potential wells.* In the case of the harmonic oscillator, the well is parabolic in shape. In the case of the particle in a box, the well is a square well; it has a flat bottom from which the potential energy rises abruptly to infinity at the ends of the box.

You will notice an important difference between the two cases, however, when you remember what the particle in the parabolic well would do if it were behaving classically. If the particle had the total energy E_0, for example, it would oscillate back and forth along the line E_0 in Fig. 5.1, between the limits A_0 and $-A_0$ representing the amplitude of the oscillation. At the points A_0 and $-A_0$ it would come instantaneously to rest, and all its energy would be in the form of potential energy, as the intersection of the line for E_0 with the curve for U_{pot} at those points makes clear.

But the wave functions for the quantum-mechanical oscillator assign a definite probability that the particle will be beyond those points. In those 'classically forbidden' regions, clearly the total energy of the particle would be less than its potential energy, and thus its kinetic energy would be *negative* there. This remarkable result has come from the fact that the potential energy for the oscillator is not infinite outside the classically permitted region, as it is in the case of the particle in a box.

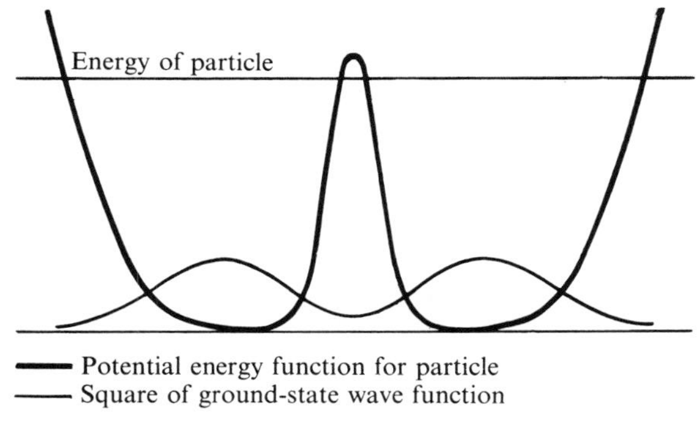

FIG. 5.2. A particle with too little energy to traverse a barrier, according to classical mechanics, may leak through the barrier, according to quantum mechanics.

The suggestion that a particle can be found in a region where classical mechanics would not permit it to penetrate is an important practical result of quantum mechanics. Sometimes the potential for a particle will have two or more wells, separated by a thin region of high potential energy, as in Fig. 5.2. The particle then has a finite chance of traversing the barrier, even if its total energy is small. The phenomenon is sometimes called *barrier penetration*, or the *tunnel effect*.

Compare now the wave-mechanical probability distribution for the harmonic oscillator with the classical distribution. The probability that the particle is in a certain range of x is proportional to the time spent in that range. Hence the classical probability is inversely proportional to the velocity of the particle in that range. Clearly then it will be largest near $x = \pm A$ and smallest near $x = 0$.

Comparing that probability with the square of a wave function of high quantum number, as in Fig. 5.3, you see that the quantum prediction is oscillating about the classical prediction, and shows n 'zeros' along the x axis. Averaged over ranges greater than the distance between these zeros, the quantum probability would approach the classical probability. That

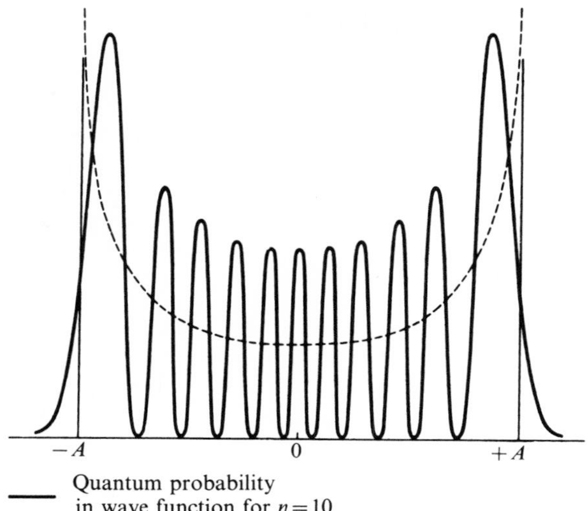

FIG. 5.3. The probability function for a harmonic oscillator in a state of high quantum number oscillates about the classical probability.

reflects a general feature of quantum mechanics: in the limit of high quantum numbers, quantum mechanics approaches classical mechanics. It is in this sense that quantum mechanics contains classical mechanics instead of conflicting with it.

PROBLEMS

5.1 Derive the classical probability function for the harmonic oscillator shown in Fig. 5.3.

5.2 In the limit of high quantum numbers, what is the approximate distance within which the quantum probability function for the harmonic oscillator (Fig. 5.3) performs one oscillation, in terms of the mass, the classical frequency, and the classical amplitude?

5.3 Using the scheme of eqn (5.4), calculate the expectation value of the kinetic energy for the ground state of the harmonic oscillator. What is the ratio of mean kinetic to total energy? Is it the same as in classical mechanics? The integrals you will need are of the form

$$\int_{-\infty}^{\infty} \exp(-ax^2)x^{2k}\,dx = \frac{1\,.\,3\ldots(2k-1)}{2^k}\sqrt{\left(\frac{\pi}{a^{2k+1}}\right)}.$$

It can be shown that this ratio is the same in *all* the states of the harmonic oscillator.

5.4 Use the probability interpretation of the wave function to discuss two aspects of the behaviour of a particle in the immediate neighbourhood of a place where its wave function has a node.
(a) What is the probability of finding the particle there?
(b) Is the kinetic energy finite, infinite, or zero there?

6. Orthogonality and Kinetic Energy

WAVE functions for all sorts of mechanical situations have several noteworthy properties in common. Those properties may seem purely mathematical at first; their physical significances will become clearer as you use them in practical cases.

One such property is the *orthogonality* of the various wave functions appropriate to any one physical problem. Two functions of x, ψ_n and ψ_m, are said to be *orthogonal on the interval* a *to* b if

$$\int_a^b \psi_m \psi_n \, \mathrm{d}x = 0.$$

The property of orthogonality may be familiar to you already in connection with Fourier series. You see it exemplified again, in exactly that form, in the wave functions for the particle in a box. There, as Chapter 3 pointed out, the wave functions are $\Psi_n = A_n \sin(n\pi x/l)\exp -2\pi i\nu t$ in the range of x from 0 to l, and vanish elsewhere. The quantum number n is an integer, and thus the space-dependent parts of any pair of these wave functions ψ_n and ψ_m, where $n \neq m$, are orthogonal on the interval 0 to l. Recalling that the total energy associated with a state is also determined by n, you see that *the wave functions associated with states of different energies are orthogonal.*

A proof of that statement can be written quite simply for cases in which one particle is moving in one dimension, in a potential having any form $U_{\mathrm{pot}}(x)$ that leads to a discrete set of stationary energy states. The particle on a circle (Chapter 2), the particle in a box (Chapter 3), and the harmonic oscillator (Chapter 5) all provide examples of this general form. When the wave functions are complex, the definition of orthogonality just given is extended so that it reads

$$\int_a^b \psi_m^* \psi_n \, \mathrm{d}x = 0. \tag{6.1}$$

Schrödinger's equation for such cases is

$$\frac{-h^2}{8\pi^2 m}\frac{\mathrm{d}^2\psi_n}{\mathrm{d}x^2} + U_{\mathrm{pot}}(x) \, . \, \psi_n = E_n \psi_n. \tag{6.2}$$

Since this equation is linear and real, ψ_m^* is a solution for the same value of E as ψ_m; in other words

$$\frac{-h^2}{8\pi^2 m}\frac{\mathrm{d}^2\psi_m^*}{\mathrm{d}x^2} + U_{\mathrm{pot}}(x) \, . \, \psi_m^* = E_m \psi_m^*. \tag{6.3}$$

The first step in the proof is to eliminate U_{pot} by multiplying eqn (6.2) by ψ_m^* and eqn (6.3) by ψ_n, and subtracting one from the other:

$$\frac{-h^2}{8\pi^2 m}\left(\psi_m^*\frac{\mathrm{d}^2\psi_n}{\mathrm{d}x^2}-\psi_n\frac{\mathrm{d}^2\psi_m^*}{\mathrm{d}x^2}\right) = (E_n-E_m)\psi_m^*\psi_n. \tag{6.4}$$

Since the bracketed part of the left side of eqn (6.4) equals

$$\frac{\mathrm{d}}{\mathrm{d}x}\left(\psi_m^*\frac{\mathrm{d}\psi_n}{\mathrm{d}x}-\psi_n\frac{\mathrm{d}\psi_m^*}{\mathrm{d}x}\right),$$

multiply both sides of the equation by $\mathrm{d}x$ and integrate, to obtain

$$\frac{-h^2}{8\pi^2 m}\left|\psi_m^*\frac{\mathrm{d}\psi_n}{\mathrm{d}x}-\psi_n\frac{\mathrm{d}\psi_m^*}{\mathrm{d}x}\right|_a^b = (E_n-E_m)\int_a^b \psi_m^*\psi_n\,\mathrm{d}x. \tag{6.5}$$

Choosing for the range a to b the entire region accessible to the particle (which usually extends to infinity in both directions), you know that the left side must vanish at the limits because ψ_m^* and ψ_n vanish there. Hence

$$(E_n-E_m)\int_a^b \psi_m^*\psi_n\,\mathrm{d}x = 0. \tag{6.6}$$

This equation shows that when E_n is not equal to E_m the integral must vanish, and hence the functions ψ_m and ψ_n are orthogonal.

Many important mechanical situations arise in which $E_n = E_m$ but $\psi_m \neq \psi_n$. These situations, where several wave functions correspond to the same energy, are called *degenerate*. You have already seen a case of two-fold degeneracy in the particle on a circle: two distinct wave functions are solutions of Schrödinger's equation with the same value of E. Since Schrödinger's equation is linear, any linear combination of those solutions is also a solution with the same value of E. By taking such linear combinations you can make, out of any set of degenerate wave functions, another set of an equal number of wave functions which are orthogonal and which satisfy Schrödinger's equation with the same value of the energy. The technique for doing this, in a case of twofold degeneracy whose wave functions are real, is shown in Discussion 6.1. The procedure is easily extended to handle degeneracy of any degree.

Some feeling for the physical meaning of the orthogonality of the different wave functions appropriate to any particular mechanical situation can be obtained in the following way. Think first of a classical mechanical case in which you have already encountered orthogonal functions: the description of the normal modes of vibration of a stretched string by the space functions $f_n(x) = A_n \sin(n\pi x/l)$. There you could imagine exciting any one normal mode independently of the rest, and could find that the mode persists and does not feed energy into other modes. In other words, the modes are not coupled.

Discussion 6.1

ORTHOGONALIZING DEGENERATE WAVE FUNCTIONS

Suppose that ψ_a and ψ_b are a pair of degenerate wave functions that are not orthogonal. You want to construct a wave function ψ_c which is a linear combination of both of them and which is orthogonal to one of them, say to ψ_a. Setting

$$\psi_c = a\psi_a + b\psi_b,$$

where a and b are constants whose values are to be determined, you require

$$\int \psi_a \psi_c \, d\tau = \int \psi_a \cdot (a\psi_a + b\psi_b) \, d\tau = 0.$$

If ψ_a and ψ_b have been normalized, you know that $\int \psi_a^2 \, d\tau = 1$, and the foregoing equation becomes

$$a + b\int \psi_a \psi_b \, d\tau = 0.$$

An additional equation connecting a and b arises when you normalize the newly constructed wave function ψ_c:

$$\int (a\psi_a + b\psi_b)^2 \, d\tau = a^2 + b^2 + 2ab\int \psi_a \psi_b \, d\tau = 1.$$

When $\int \psi_a \psi_b \, d\tau$ has been evaluated, the two equations provide two algebraic relations for evaluating the coefficients a and b, and thus exhibiting a pair of normalized wave functions ψ_a and ψ_c, still degenerate but now orthogonal.

This procedure, here demonstrated for wave functions that are real, is easily generalized for complex wave functions.

That sort of independence—that lack of coupling—is the feature reflected in the orthogonality of the functions that describe the modes.

The orthogonality of wave functions has an analogous physical interpretation. Associated with each wave function is an exact value of the energy, and you find that the wave function describes a stationary energy state, unchanging with time and uncoupled to other possible states of the system. It is fruitful to pursue this analogy into cases where coupling between the normal modes is introduced, and several words used in discussing chemical bonds, such as the word 'resonance', are derived from the analogy.†

Notice one feature of wave functions which qualitatively connects their orthogonality with their kinetic energies. In the case of the particle in a box (Fig. 3.1) you can see especially clearly how the orthogonality of the first two wave functions arises. The second function, multiplying the first function, gives an integrand in which there are parts which are the exact negatives of other parts all through the range of integration, so that the whole integral vanishes. Since the first function is positive throughout the range, the second function can accomplish orthogonality only by being positive in some places and negative in others, or in short by passing through zero at least once in

† This analogy is discussed further in *Bonds between atoms*, Chapter 8.

the range. Pursuing this observation to higher energies, you see that the higher the energy of the wave function the larger the number of zeros it has.

Now clearly the larger the number of its zeros in a given range, the more the function oscillates in that range. And the more it oscillates, the larger the value of its second derivative is likely to be—the more rapidly its slope must change. But the larger the second derivative, the higher the kinetic energy, as you can see from the way Schrödinger's equation was constructed in Chapter 4.

The wave functions for the harmonic oscillator show these features clearly. States of high quantum number (Fig. 5.3) have their highest density of zeros in the neighbourhood of $x = 0$, where the particle moves fastest. Even the ground-state wave function (Fig. 5.1) shows its greatest curvature at $x = 0$.

It is worthwhile pursuing the harmonic oscillator a step further, in order to illustrate how you might make quantitative application of these ideas. Suppose you set out to estimate the mean kinetic energy of a harmonic oscillator by regarding each of its states as if it were that of a particle in a one-dimensional box, and using a different size of box for each state. Evidently an appropriate choice for the size of the box is twice the classical amplitude corresponding to the state. Using the foregoing argument about oscillations, you would choose for the wave function in the box that which has the same number of zeros as the true wave function for the state. You could describe your procedure as approximating to the mean kinetic energy by making

Discussion 6.2

ESTIMATING THE KINETIC ENERGY OF A HARMONIC OSCILLATOR

Pursuing the suggestion of the text, you would find the box size for each state by equating the classical energy to the quantum mechanical energy (eqns 5.14 and 5.19):

$$(n+\tfrac{1}{2})h\nu = \tfrac{1}{2}kA^2,$$

After examining Fig. 5.1, you would pick the de Broglie wavelength of your approximating wave functions as

$$\lambda = \frac{4A}{n+1}.$$

The approximate mean kinetic energy that you seek will be given by

$$U_{\text{kin}} = \frac{p^2}{2m} = \frac{h^2}{2m\lambda^2}.$$

Eliminating A and λ between these three equations, and using the relation (5.8) between ν, k, and m for the harmonic oscillator, you obtain

$$U_{\text{kin}} = \frac{\pi^2}{8}h\nu\frac{(n+1)^2}{2n+1}.$$

a rational choice of de Broglie wavelength. The mathematical details, carried out in Discussion 6.2, lead to the estimation

$$U_{\mathrm{kin}} = \frac{\pi^2}{8} h\nu \frac{(n+1)^2}{2n+1}, \tag{6.7}$$

which for the ground state is $(\pi^2/8)h\nu$, and for large n approaches $(\pi^2/16)nh\nu$.

At once you have no doubt that the value for the ground state is wrong. It provides a kinetic energy more than twice as large as the total energy (eqn 5.15), and that is impossible because the potential energy is positive. The approximation for large n is much closer to the truth, but it is still too high. Clearly the reason for these errors comes from the fact that the true wave function extends outside the box; the particle is not confined to its classical province and its effective de Broglie wavelength is somewhat longer than that which you have picked. But the approximation is of the right order of magnitude at its worst, gives nearly the right dependence of U_{kin} on the quantum number, n, and exhibits the right dependence on the frequency ν.

PROBLEMS

6.1 Accepting the fact that the wave functions for the harmonic oscillator will be of the form $\psi_n = \exp -\alpha x^2/2(a_0 + \ldots + a_n x^n)$, show schematically how you can determine the a's in these wave functions without using the wave equation, by starting with ψ_0 and requiring each successive wave function to be orthogonal to those you have already constructed.

6.2 Using the result of Problem 5.3, and the method of approximation employed in this chapter, show that the true effective de Broglie wavelength of the harmonic oscillator is

$$\lambda = 2\pi\sqrt{2\frac{A}{2n+1}}.$$

7. The Variational Method of Approximation

A USEFUL application of the orthogonality of the various wave functions appropriate to a physical problem, discussed in the preceding chapter, is the variational method of approximation. It is a method for calculating an approximate value of the lowest energy of a physical system when the problem of finding that energy cannot be solved exactly. The method can be adapted to obtain approximations to the energies of other states, but the state of lowest energy—the ground state—is of most frequent interest, since any system will adopt that state unless it is excited.

For simplicity, consider a physical system whose time-independent wave functions are real. If you had been able to find the true wave functions, you could exhibit a set of solutions to the time-independent Schrödinger equation, and an associated set of energy levels,

$$\begin{gathered} \psi_0, \psi_1, \dots \psi_n, \dots ; \\ E_0, E_1, \dots E_n, \dots . \end{gathered} \tag{7.1}$$

From the arguments of the preceding chapter you know that the wave functions would be orthogonal on an appropriate interval:

$$\int_b^a \psi_m \psi_n \,\mathrm{d}\tau = 0, \quad \text{when} \quad m \neq n. \tag{7.2}$$

And you know that you could choose a coefficient for each wave function so that it would be normalized:

$$\int \psi_n^2 \,\mathrm{d}\tau = 1. \tag{7.3}$$

Now suppose that you are prepared to guess the general form of the wave function for the ground state. Call that guess ϕ. Unless you happen to have guessed exactly right, ϕ will not be a solution to the time-independent Schrödinger equation for the system. In other words, if you found some way of placing the system in the state described by ϕ, the system would not remain in that state, with a definite value of the energy; its state would change with time. But you could calculate an *expectation value* of the energy in that state as Discussion 5.1 described:

$$\overline{H} = \frac{\int \phi H \phi \,\mathrm{d}\tau}{\int \phi^2 \,\mathrm{d}\tau}, \tag{7.4}$$

where H is the true Hamiltonian operator for the problem. The question then arises, what is the relationship of $\bar{H}$ to any of the true E's for the system?

The relationship can be found by noticing that, if you knew the true wave functions—the ψ's of (7.1)—you could expand ϕ in a series of those ψ's. The method of expansion is a generalization of the familiar method by which any periodic function can be represented as a Fourier series. Examine a proposed expansion,

$$\phi = a_0\psi_0+a_1\psi_1+\ldots+a_n\psi_n+\ldots \tag{7.5}$$

where the a's are coefficients to be determined. Multiply both sides of (7.5) by ψ_n and integrate over the interval on which the ψ's are orthogonal. Then

$$\int \psi_n\phi \, \mathrm{d}t = \psi_n \cdot (a_0\psi_0+a_1\psi_1+\ldots+a_n\psi_n+\ldots) \, \mathrm{d}\tau. \tag{7.6}$$

But all terms on the right vanish except the nth term because the functions ψ are orthogonal, and the nth term equals a_n since the functions ψ have been normalized. Hence each of the coefficients can be found by evaluating

$$a_n = \int \psi_n\phi \, \mathrm{d}\tau. \tag{7.7}$$

In the present case you do not know the ψ's, but you can imagine a substitution of (7.5) into (7.4), yielding

$$\bar{H} = \frac{\int (a_0\psi_0+\ldots)H(a_0\psi_0+\ldots) \, \mathrm{d}\tau}{\int (a_0\psi_0+\ldots)^2 \, \mathrm{d}\tau}. \tag{7.8}$$

And now you notice that, since the ψ's are true solutions to the Schrödinger equation for the problem,

$$H\psi_n = E_n\psi_n, \tag{7.9}$$

and therefore

$$H(a_0\psi_0+\ldots) = a_0E_0\psi_0+a_1E_1\psi_1+\ldots\ . \tag{7.10}$$

Eqn (7.8) becomes

$$\bar{H} = \frac{\int (a_0\psi_0+\ldots)(a_0E_0\psi_0+\ldots) \, \mathrm{d}\tau}{\int (a_0\psi_0+\ldots)^2 \, \mathrm{d}\tau}, \tag{7.11}$$

and since the ψ's are orthogonal and are normalized,

$$\bar{H} = \frac{a_0^2E_0+a_1^2E_1+\ldots}{a_0^2+a_1^2+\ldots}. \tag{7.12}$$

If the numerator in eqn (7.12) were $a_0^2E_0+a_1^2E_0+\ldots$, then $\bar{H}$ would equal E_0. But the numerator is greater than that because, if E_0 is the energy of the ground state, all other E's are greater. Thus you finally infer that $\bar{H}$ calculated by eqn (7.4) will always be greater than the true ground-state energy.

This may seem like a modest achievement, but you will respect it more when you examine some of its implications. In the first place, you can always establish an upper limit to the energy of the ground state, and that can be quite a useful thing to do. In the second place, you can choose for ϕ a function embodying some variable parameters. The $\bar{H}$ calculated from eqn (7.4) will be a function of those parameters. Then you can minimize $\bar{H}$ with respect to variations in those parameters, with the assurance that the minimum $\bar{H}$ will still be greater than the true energy. Using a shrewdly chosen *variation*

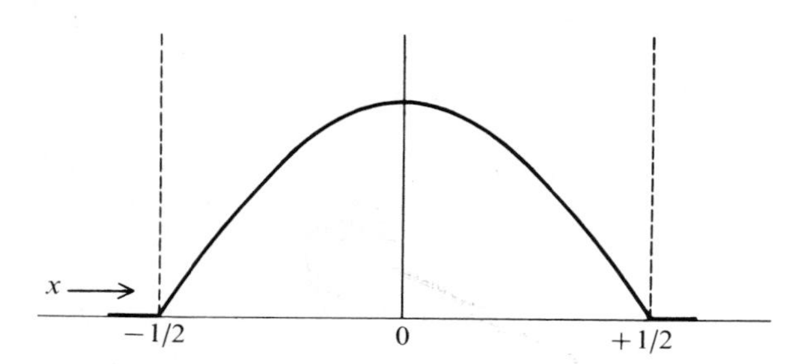

FIG. 7.1. Variation function for the ground state of the particle in a one-dimensional box.

function ϕ, which contains several undetermined parameters, you can sometimes get a very close approximation to the true energy.

As an example of how well this procedure may serve, even without a parameter to vary, consider the problem of the particle in a box of width l with infinitely high walls. For this problem the discussion in Chapter 3 has provided a true value of the ground-state energy, and you can compare that value with an approximate value in order to see how well the method of approximation works. Your approximating effort might proceed as follows.

From the discussions of particles in potential wells, in preceding chapters, you know that the wave function for the ground state must reach a maximum at the middle of the box, go to zero at the sides of the box, and vanish everywhere outside the box. Taking the sides of the box at $x = \pm(l/2)$, you could try the function (Fig. 7.1)

$$\phi = \frac{l^2}{4} - x^2 \text{ inside,} \qquad \phi = 0 \text{ outside.} \tag{7.13}$$

The true Hamiltonian operator for this problem is simply

$$H = \frac{-h^2}{8\pi^2 m}\frac{d^2}{dx^2}, \tag{7.14}$$

since the particle is free inside the box. Putting (7.13) and (7.14) into eqn (7.4), you obtain

$$\overline{H} = \frac{\displaystyle\int_{-l/2}^{+l/2} \frac{h^2}{4\pi^2 m}\left(\frac{l^2}{4} - x^2\right) dx}{\displaystyle\int_{-l/2}^{+l/2} \left(\frac{l^2}{4} - x^2\right)^2 dx} = \frac{5h^2}{4\pi^2 m l^2}. \tag{7.15}$$

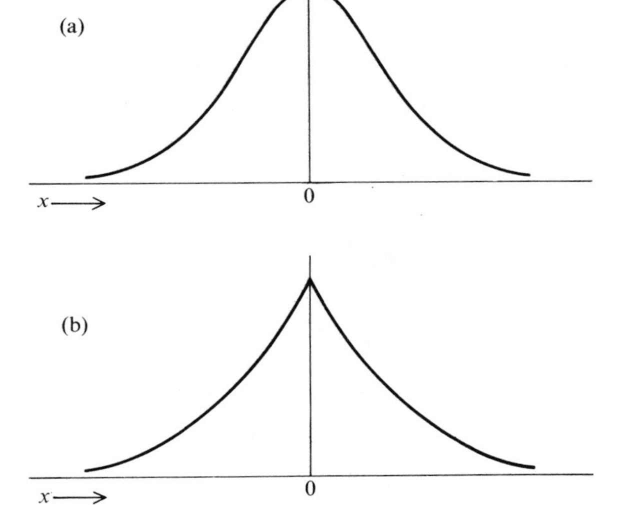

FIG. 7.2. The ground state (a) of the harmonic oscillator, and a variation function (b) for that state.

But for this problem you know that the true energy levels are $n^2h^2/8ml^2$, and thus that the ground-state level ($n = 1$) is $h^2/8ml^2$. Notice that, if π^2 were equal to 10, your answer (7.15) would be exact. Actually, since π^2 is a little less than 10, your approximate value is a little greater than the true value, as the preceding discussion of the variation approximation would lead you to expect.

For another example, take the problem of the simple harmonic oscillator. Here again the wave function for the ground state must have a maximum in the middle, and it must decline in both directions from there, approaching zero at an infinite distance. You might try the variation function (Fig. 7.2b):

$$\begin{aligned} \phi &= \mathrm{e}^{ax} \quad \text{for} \quad x \le 0; \\ \phi &= \mathrm{e}^{-ax} \quad \text{for} \quad x \ge 0. \end{aligned} \tag{7.16}$$

The Hamiltonian operator for the problem is

$$H = \frac{-h^2}{8\pi^2 m}\frac{d^2}{dx^2}+\tfrac{1}{2}kx^2. \tag{7.17}$$

Then from (7.4) you get

$$\bar{H} = \frac{2\int_0^\infty\left(\frac{-h^2a^2}{8\pi^2 m}+\frac{1}{2}kx^2\right)e^{-2ax}\,dx-\frac{h^2}{8\pi^2 m}\int_{x=0}\phi\frac{d^2\phi}{dx^2}\,dx}{2\int_0^\infty e^{-2ax}\,dx}. \tag{7.18}$$

Here the last term in the numerator needs explanation. The function (7.16) has a discontinuity of slope at $x = 0$; the function turns infinitely rapidly there over an infinitesimally short distance. Since the second derivative is infinite at that point, the 'corner' makes a finite contribution to the numerator. That contribution can be evaluated by noticing that

$$\int_{x=0}\phi\frac{d^2\phi}{dx^2}\,dx = \phi_{x=0}\int_{x=0} d\,\frac{d\phi}{dx} = \phi_{x=0}\left(\left.\frac{d\phi}{dx}\right|_{\text{right}}-\left.\frac{d\phi}{dx}\right|_{\text{left}}\right) = -2a. \tag{7.19}$$

Hence you obtain

$$\bar{H} = \frac{\frac{-h^2a^2}{8\pi^2 m}\cdot\frac{1}{2a}+\frac{k}{8a^2}+\frac{h^2a}{8\pi^2 m}}{\frac{1}{2a}}$$

$$= \frac{h^2a^2}{8\pi^2 m}+\frac{k}{4a^2}. \tag{7.20}$$

In order to minimize $\bar{H}$ with respect to variation of the parameter a, set

$$\frac{\partial\bar{H}}{\partial a} = \frac{h^2a}{4\pi^2 m}-\frac{k}{2a^3} = 0,$$

whence

$$a^4 = \frac{2\pi^2 mk}{h^2}. \tag{7.21}$$

Then the best approximation with this variation function gives

$$\bar{H} = \frac{\sqrt{2}h}{4\pi}\sqrt{\left(\frac{k}{m}\right)} = \frac{\sqrt{2}}{2}h\nu, \tag{7.22}$$

which is higher than the true ground-state energy (eqn 5.17) by the factor $\sqrt{2}$.

Thus this approximation is somewhat less successful than the approximation to the ground-state energy of the particle in a box, even though you have taken advantage of a variable parameter. You might have expected this, for the variation function has not been as well chosen: it looks less like the true wave function. In general the more nearly the chosen functional form resembles the true wave function, the better is the result of a variational calculation of the energy.

Note well, however, that the converse of that last remark is not necessarily true. The variational theorem does not say that the more nearly correct the energy is, the more nearly correct the wave function is. It is easy to see why that need not be true. The approximate energy is calculated by performing integrations, and there are infinitely many functions whose integrals over a single range would have the same value.

PROBLEMS

7.1 You have noticed that the wave functions for a particle in a symmetrical one-dimensional well are alternately symmetric and antisymmetric about the centre of the well, as the energy of the particle is increased in successive steps. For such a particle you might expect that $\bar{H}$ calculated by using an antisymmetric variation function would give an approximation to the energy of the first excited state—the state of lowest energy that such a function is capable of approximating. Check this expectation, for the particle in a box of width l (Fig. 7.3), by using for the trial function

$$\phi = \frac{-l^2}{16}+\left(x+\frac{l}{4}\right)^2 \quad \text{for} \quad -\frac{l}{2} \le x \le 0,$$

$$\phi = \frac{l^2}{16}-\left(x-\frac{l}{4}\right)^2 \quad \text{for} \quad 0 \le x \le \frac{l}{2},$$

$$\phi = 0 \text{ elsewhere.}$$

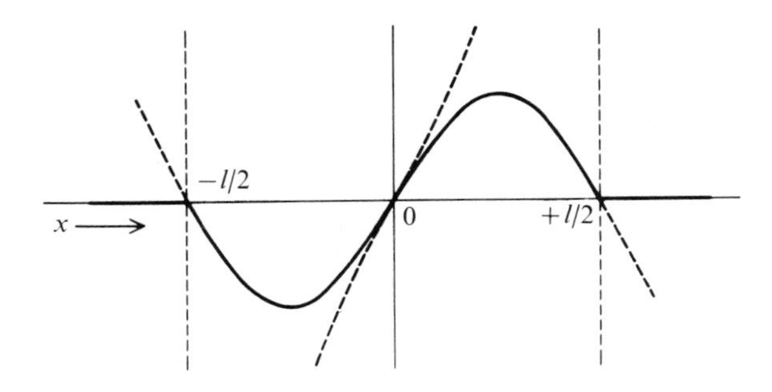

FIG. 7.3.

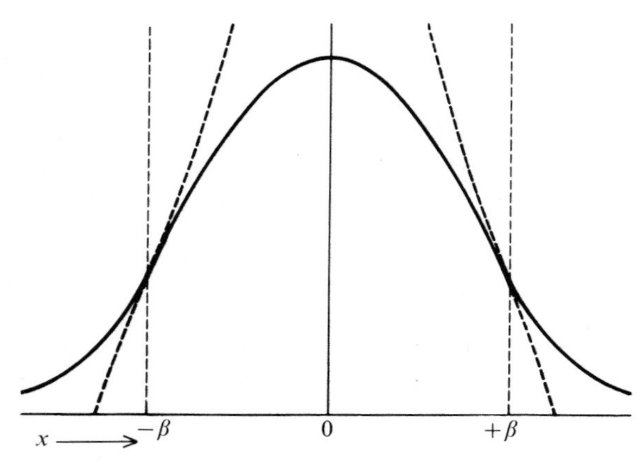

FIG. 7.4.

7.2 Find a better approximation to the ground-state energy of the harmonic oscillator than that given by the variation function (7.16) by using a variation function which has the form

$$\phi = a - bx^2 \quad \text{for} \quad -\beta \leq x \leq \beta,$$

$$\phi = \frac{c}{x^2} \text{ elsewhere,}$$

choosing the coefficients a, b, c so that the function and its first derivative match at $x = \pm\beta$, and minimizing $\bar{H}$ with respect to the parameter β (Fig. 7.4).

8. The Spinning Electron

THE electron is a 'fundamental particle'—that is to say, we cannot at present ascribe to it a *structure* that makes it appear to be built of other particles, as an atom is built of a nucleus surrounded by electrons, or as a nucleus is built of protons and neutrons. Much of the behaviour of an electron, when it is in a stationary state, can therefore be described quite well by the wave-mechanical methods that have been developed for describing particles, discussed in preceding chapters. The mass of an electron can be taken to be $m = 9{\cdot}1 \times 10^{-31}$ kg, and its electrostatic charge can be taken to be $e = -1{\cdot}6 \times 10^{-19}$ C.

In order to explain some aspects of its behaviour, however, a third property must be ascribed to it besides its mass and its charge. That property is ordinarily called its *spin*. Accordingly the *state* of an electron is not completely defined by such a wave function as the preceding chapters have described: you must pay attention also to its spin. The name spin is appropriate for two reasons. In the first place, there are observable phenomena in which you can think of this spin as changing direction. Then what happens can be explained by ascribing an angular momentum to the spinning electron and using the laws of conservation of angular momentum—laws that have been found applicable everywhere else in classical and quantum mechanics.† In the second place, when the spin changes direction, there is a change of magnetic moment, as if the direction of current in a little current loop had reversed.

If you want to think of the electron as a charged ball, think of the ball as spinning around an axis through its centre. There will be occasions when you must think of the ball as occupying the same place as the nucleus of an atom, however; a wave function may give it a finite probability of being found precisely at the nucleus. If the idea of two balls in one and the same place does violence to your picture, think of the electron as a torus, or a little current loop that runs round the nucleus while the electron is at it.

But better yet, here as elsewhere in quantum mechanics, practise the art of switching pictures, and even perhaps wiping them all out. Indeed, if you persist in thinking of the electron as a charged spinning ball, in such a concrete fashion that you calculate the magnetic moment of the ball from its charge and its angular momentum, your answers will probably not agree with experiment.‡

† In quantum mechanics those laws apply to the *expectation value* of angular momentum.

‡ It is a fact—but probably a not very significant fact—that the observed ratio of magnetic moment to angular momentum, $-e/m$, is the ratio appropriate to a spherical ball, whose mass m is uniformly distributed through its *bulk*, and whose charge $-e$ is uniformly distributed over its *surface*, rotating about an axis through its centre.

As with any spinning object, the only way to describe the direction of spin is to establish some axial system to which the direction can be referred. For an isolated spinning object, nature provides no axial system. One axis can be provided by applying a magnetic field, and that is an axis to which the electron is responsive because it has a magnetic moment. It tends to align its magnetic moment, and thus its spin, with the field.

And when it does so, experiments show that its spin is quantized so rigorously that the components of angular momentum and of magnetic moment along the field can take only two different values, equal in magnitude and opposite in direction. The possible values of that component of angular momentum are $\pm(h/4\pi)$ and of the corresponding component of magnetic moment are $\mp(eh/4\pi m)$. The latter quantity is often used as a unit of magnetic moment, the *Bohr magneton*, with the value $9{\cdot}27\times10^{-24}$ J T^{-1}.

Now of course angular momenta and magnetic moments are vector quantities, as Discussion 8.1 will remind you. When their z components

Discussion 8.1

ANGULAR MOMENTA IN QUANTUM MECHANICS

The angular momentum of a system of particles about a point is a quantity that is useful in describing the behaviour of a system by either classical or quantum mechanics. It is a vector quantity, equal to the vector sum of the angular momenta of each constituent particle. And for each particle the angular momentum is defined in classical mechanics as the vector product,

$$\mathbf{M} = \mathbf{r} \times \mathbf{p},$$

where $\mathbf{r}$ is the position vector from the chosen point to the particle, and $\mathbf{p}$ is the linear momentum of the particle.

When the Cartesian components of this vector are written,

$$M_x = yp_z - zp_y,$$
$$M_y = zp_x - xp_z,$$
$$M_z = xp_y - yp_x,$$

the appropriate quantum-mechanical operators for these components can be written by substituting for the p's the operators for the components of linear momentum described in Chapter 5:

$$M_x = \frac{h}{2\pi i}\left(y\frac{\partial}{\partial z} - z\frac{\partial}{\partial y}\right),$$

$$M_y = \frac{h}{2\pi i}\left(z\frac{\partial}{\partial x} - x\frac{\partial}{\partial z}\right),$$

$$M_z = \frac{h}{2\pi i}\left(x\frac{\partial}{\partial y} - y\frac{\partial}{\partial x}\right).$$

By using these operators, expectation values and the like can be calculated as described for operators in general in Chapter 5.

have been determined, it is natural to inquire into the other two components of the angular momentum and magnetic moment of the electron. In order to give meaning to those components, the physical system must be endowed with further objects or fields of force, sufficient to specify a three-dimensional axial system, perhaps by adding nuclei and applying an electric field. But after those things have been said and done, you will encounter a severe restriction imposed by the uncertainty principle.

The discussion in Chapter 3 connected uncertainties in a coordinate and in the momentum along that coordinate. Angular momenta could not come into question in the one-dimensional systems that the preceding chapters have described. But in such phenomena as the behaviour of electrons in atoms, the angular momentum of the system of particles about its centre of mass is as important for the quantum-mechanical description as for the classical.

The uncertainties connecting the expectation values of the components of angular momentum of a system are discussed in the appendix to this chapter. In particular the root-mean-square uncertainties in M_x and M_y, the x component and the y component of the angular momentum of a mechanical system, are connected by

$$(\Delta M_x)(\Delta M_y) \geq \frac{h}{4\pi}|\overline{M}_z|, \tag{8.1}$$

where $|\overline{M}_z|$ is the absolute expectation value of the z component of angular momentum. And of course there are two other relations obtained by interchanging x, y, and z.

These results, obtained in the first place for systems of particles moving in three-dimensional space, may seem irrelevant to the spinning electron. But it is always appropriate to test a result, obtained in one context, for applicability in another. Substituting into the relation (8.1) the value $|M_z| = h/4\pi$ already mentioned, you find that

$$(\Delta M_x)(\Delta M_y) \geq \left(\frac{h}{4\pi}\right)^2, \tag{8.2}$$

for the spin of the electron. There is no reason to suppose that ΔM_x, would differ from ΔM_y, and you conclude that

$$\Delta M_x = \Delta M_y \geq \frac{h}{4\pi} = |\overline{M}_z|. \tag{8.3}$$

Accepting this result would force the conclusion that there is a *complete* uncertainty in the values of M_x and M_y. Surely the values of those components cannot be larger than the value along the preferred axis z. And the behaviour of the spinning electron turns out to be experimentally consistent with this result. When you have established one axis, and determined which of the two possible values of spin the electron has relative to that axis, you have done and said all that can be done and said about the spin of the electron.

The conventional method of specifying the spin quantum state of an electron is suggested by a comparison with more familiar mechanical behaviour. In many systems of particles—in a hydrogen atom, for example—the component of angular momentum along an axis is also quantized, changing in multiples of $h/2\pi$, the *Bohr unit of angular momentum.* For the spin of an electron, the corresponding component can take the values $\pm(h/4\pi)$. It is therefore convenient to specify the spin quantum state of an electron by a spin quantum number m_s, which can take the values $m_s = \pm\frac{1}{2}$.

Now when a system contains more than one electron in reasonably close association, you think of the magnetic field due to the magnetic moment of any one of them as establishing a z axis for the rest. In the presence of that electron, the remaining electrons have m_s values of $\pm\frac{1}{2}$. Hence, regardless of which electron you pick to establish the z axis, the spins of two or more electrons will be lined up parallel or antiparallel. If there is nothing outside the system that establishes a z axis, the electrons must correlate their spin axes nevertheless; there will be a set of electrons with $m_s = +\frac{1}{2}$ and a set with $m_s = -\frac{1}{2}$. In speaking of such systems physicists and chemists often refer to the electrons with spin 'up' and those with spin 'down'—convenient monosyllables to describe the two antiparallel groups. The state of a system of electrons is completely defined, therefore, by a wave function of the coordinates of all the electrons and the time, together with a specification of all the spin quantum numbers.

The formal methods of quantum mechanics have been extended to accommodate the property of spin, but a thorough discussion of that extension would carry us too far afield. Suffice it to say that you can think of spin as introducing a new coordinate, or set of coordinates, for the electron. These coordinates become new observable quantities, having appropriate *operators.* The wave function for an electron becomes a function not only of the familiar space coordinates but also of the new spin coordinates. Rules for using the spin operators on these enlarged wave functions, analogous to those described for more familiar properties in preceding chapters, enable you to calculate expectation values and the like. For very many cases the formalism permits a *separation of variables*, analogous to the separation of space and time variations in Chapter 4, so that the total wave function becomes a product of a function of the spin variables by a function of the space and time variables, and then you can think of the two sorts of variables separately.

More important than formalism, however, are three new features introduced into the physical picture of matter by the presence of spinning particles, with associated angular momenta and magnetic moments. Two of the new features are easy to see. In the first place, the angular momenta of the spins contribute to the total angular momentum of the system to which they belong. This contribution will affect your thinking about transitions of the system from one state to another in which the expectation value of the total angular momentum must be conserved.

In the second place, the magnetic moment associated with each spinning electron will interact with magnetic fields. Such magnetic fields may be applied from outside the system, of course. But even when no such fields are applied, magnetic forces arise from the motions of charged particles within the system, and from the magnetic moments associated with their spins. The analysis of these two new features provides an elegant description of the *multiplet* structure of atomic spectral lines, and the behaviour of those lines when magnetic forces are deliberately applied.

The third new feature, though unexpected, is probably the most important. The spin coordinates must be included in the applications of a law of nature which has no classical counterpart. This *law of antisymmetry* leads to the 'Pauli exclusion principle'—a principle of the utmost importance in explaining atomic, molecular, and solid behaviour—in a way which the next chapter will describe

APPENDIX

UNCERTAINTIES IN ANGULAR MOMENTA

Consider two of the many possible meanings that could be ascribed to *the uncertainty* in something. It might mean, in the first place, the *total range* over which something could vary without permitting a determination of its value more precisely. It was feasible to use that meaning in speaking of the particle in a box in Chapter 3 because the range of position of the particle is precisely limited by the walls of the box. But clearly that meaning would give difficulty in the case of the harmonic oscillator because the wave functions all ascribe some probability that the particle is anywhere.

You can give a more generally applicable meaning to *the uncertainty* by borrowing an idea from statistical theory. There it is common to describe the variation of a quantity by the following procedure. First find the average of all the measurements of the quantity, then find all the individual deviations from that average, then square all those deviations, then add those squares together, then divide by the number of observations, and finally take the square root of the result, to obtain the *root-mean-square deviation* of the quantity. You can see some of the reason for this procedure by imagining some alternatives. Averaging the deviations themselves would yield zero, for they are as often negative as positive. Averaging the absolute values of the deviations yields a significant quantity, but mathematical difficulties afflict the interpretation of the results.

It is natural to derive a definition of *uncertainty* from the idea of root-mean-square deviation. The expectation value of an observable quantity is just the average of the values which you would get by many measurements of it, as you saw in Chapter 5. For example, an individual deviation of the coordinate x of a particle from its average would be $x-\bar{x}$, where $\bar{x}$ is the expectation value of x. The square of such a deviation would be $(x-\bar{x})^2$, the mean square would be the expectation value of that square, $\overline{(x-\bar{x})^2}$, and the root-mean-square would be the square root of that last quantity. Thus finally you can definite the uncertainty Δx in a problem by

$$\Delta x = \sqrt{\{\overline{(x-\bar{x})^2}\}}.$$

Similarly you can define the uncertainty in any other property, whose wave mechanical operator is G, by

$$\Delta G = \sqrt{\{\overline{(G-\bar{G})^2}\}}.$$

Then if the system is in a state whose normalized wave function is ψ, you find $\bar{G}$ by the procedure for finding expectation values (eqn 5.4), and then find the square of the uncertainty by applying the procedure again in accordance with the preceding equation.

The application of this definition of uncertainty to discover relations connecting the uncertainties in different quantities is straightforward, but it requires a more extensive development of the formal methods of wave mechanics than is appropriate here. When it is applied to the uncertainties in the components of angular momentum described in Discussion 8.1, the method yields the relation

$$(\Delta M_x)(\Delta M_y) \geq \frac{h}{4\pi}|\bar{M}_z|,$$

where $|\bar{M}_z|$ denotes the absolute expectation value of the z component of the angular momentum. Two more relations can be written from this one by permuting x, y, and z.

9. Symmetry and the Exclusion Principle

Two kinds of *symmetry* can appear in a wave function, which are important to recognize and to distinguish. On the one hand there can appear a symmetry in the space coordinates of a system, and on the other a symmetry to the interchange of particles of the same sort in the system.

The first kind of symmetry—spatial symmetry—is exemplified in the harmonic oscillator, described in Chapter 5. There the potential energy, $U_{pot} = \frac{1}{2}kx^2$, is symmetrical about $x = 0$: U_{pot} has the same value for $-x$ and for $+x$. Recognition of such spatial symmetries is often useful in visualizing and classifying wave functions in particular problems. You will meet many simple applications in discussing the behaviour of particles in one-dimensional wells, and the behaviour of the electrons that bond atoms together into molecules.† And the theory of the behaviour of electrons in a crystal of metal rests heavily on the fact that the atoms of a crystal are arranged in a repetitive orderly array, or in other words that their arrangement is symmetric to certain translations.

For the case of a single particle moving in one dimension, such as the harmonic oscillator, notice a consequence of symmetry about $x = 0$. Here $U_{pot}(x) = U_{pot}(-x)$. Then since $d^2/dx^2 = d^2/d(-x)^2$, the whole wave-mechanical Hamiltonian operator (Chapter 4) has this symmetry. Leaving out numerical factors for simplicity (Discussion 9.1), you can write

$$\frac{d^2\psi}{dx^2} + E\psi(x) = U_{pot}(x)\psi(x),$$

and

$$\frac{d^2\psi}{d(-x)^2} + E\psi(x) = U_{pot}(-x)\psi(x). \tag{9.1}$$

Changing x into $-x$ throughout eqn (9.1)—an entirely permissible mathematical operation—yields

$$\frac{d^2\psi(-x)}{dx^2} + E\psi(-x) = U_{pot}(x)\psi(-x). \tag{9.2}$$

Hence if $\psi(x)$ is a solution of the equation

$$\frac{d^2\psi}{dx^2} + E\psi = U_{pot}\psi, \tag{9.3}$$

† These applications are discussed in *The nature of atoms* and *Bonds between atoms*.

Discussion 9.1

ATOMIC UNITS

When Schrödinger's equation is written for electrons, it is customary to simplify the appearance of the equation by using *atomic units* of length and energy. The atomic unit of length is

$$a_0 = \frac{h^2}{4\pi^2 me^2},$$

and of energy is

$$W_{\mathrm{H}} = \frac{2\pi^2 me^4}{h^2},$$

where h is Planck's constant and e and m are the charge and mass of an electron. The atomic unit of energy (often called the Rydberg) is the negative of the energy of an electron in the ground state of a hydrogen atom, 13·53 electron volts, and the atomic unit of length is equal to the radius of the corresponding 'Bohr orbit', $0{\cdot}53 \times 10^{-8}$ cm. Then, starting with Schrödinger's equation in ordinary units for a particle moving in one dimension,

$$\frac{h^2}{8\pi^2 m}\frac{\mathrm{d}^2\psi}{\mathrm{d}x^2} + [E - V(x)]\psi = 0,$$

one makes the substitutions $E' = E/W_{\mathrm{H}}$, $x' = x/a_0$, $V'(x') = V(x)/W_{\mathrm{H}}$, and $\psi'(x') = \psi(x)$. Once one to use these units, one drops the primes, and the equation reads

$$\frac{\mathrm{d}^2\psi}{\mathrm{d}x^2} + [E - V(x)]\psi = 0.$$

Notice that the new E, x, and V are dimensionless, and that the functional form ψ is changed in such a way that its numerical value at any point in real space is unchanged by the change of units.

then $\psi(-x)$ is also a solution with the same value of the energy E. But whenever ψ is nondegenerate, $\psi(-x)$ cannot be representing any essentially different function from ψ itself. Hence you can conclude that $\psi(-x) = \pm\psi(x)$.

A function $\psi(x)$ for which $\psi(-x) = \psi(x)$ is called 'even' by mathematicians and 'symmetric to reflection' by physicists. A function for which

$$\psi(-x) = -\psi(x)$$

is called 'odd' by mathematicians and 'antisymmetric' by physicists. The result just proved is that any wave function in a nondegenerate set, belonging to a problem which is symmetric to reflection about $x = 0$, is itself either symmetric or antisymmetric to such a reflection.

You see this result exemplified in the wave functions for the harmonic oscillator shown in Fig. 5.1. There the wave functions with even quantum numbers are symmetric and those with odd quantum numbers are antisymmetric. Notice that the squares of the functions of both sorts are symmetric, in agreement with the fact that, since the system is symmetric, the observable properties of the system must be symmetric. A generalization of this result for more complicated systems is that any wave function from a nondegenerate set is either symmetric or antisymmetric to each symmetry element of the system. A wave function that is symmetric to all symmetry elements of the system is called 'totally symmetric'.

Turn now to the second kind of symmetry—symmetry to the interchange of particles of the same sort—which is even more important in quantum mechanics. It can be illustrated by considering a one-dimensional problem involving two electrons. Electrons are all alike; they are indistinguishable particles. Setting up Schrödinger's equation for a system containing two electrons, you will use two coordinates, x_1 and x_2, to describe the position of electron no. 1 and electron no. 2. But looking back in Chapter 4 at the form taken by the equation, you see that it will be unchanged if you interchange the subscripts 1 and 2. The interchange will not affect $(\nabla_1^2/m_1)+(\nabla_2^2/m_2)$ in the equation because $m_1 = m_2$ and it cannot change U_{pot} because both electrons interact with their environments and with each other in the same way when they are under the same circumstances.

Hence if $\Psi(x_1, x_2, t)$ is a solution to the equation, so is $\Psi(x_2, x_1, t)$. So also is any linear combination of the two, since Schrödinger's equation is a linear differential equation. You can construct two especially interesting linear combinations:

$$\Psi_s = \frac{1}{\sqrt{2}}[\Psi(x_1, x_2, t)+\Psi(x_2, x_1, t)],$$

and

$$\Psi_a = \frac{1}{\sqrt{2}}[\Psi(x_1, x_2, t)-\Psi(x_2, x_1, t)], \tag{9.4}$$

where $1/\sqrt{2}$ is the correct normalizing factor if Ψ was normalized in the first place. Clearly the first solution is 'symmetric' to the interchange of the subscripts, and the second is 'antisymmetric':

$$\begin{aligned} \Psi_s(x_1, x_2, t) &= \Psi_s(x_2, x_1, t), \\ \Psi_a(x_1, x_2, t) &= -\Psi_a(x_2, x_1, t). \end{aligned} \tag{9.5}$$

It can be shown that both these solutions will give symmetrical predictions for the two electrons, preserving their indistinguishability, and that no other solutions will do so. It can be shown also that each of these solutions will preserve its symmetry character as time passes, and thus that a pair of particles in a symmetric state, for example, must always remain in one. When there are more than two indistinguishable particles, the situation is more complicated, but again only the symmetric and antisymmetric states can be used.

This raises a very fundamental question about nature: are its indistinguishable particles in states that are symmetric or antisymmetric to their interchange? As always in the case of fundamental physical questions, the answer must come from experiment. That answer is embodied in the law that the most familiar fundamental material particles—electrons, protons, and neutrons—occur only in states that are antisymmetric to interchange, *when their spins are included in the specification of their state.* There are other fundamental particles—for instance photons—that occur only in symmetric states.

Return now to the wave functions for the problem involving two electrons. Add, to the specification of the state of the system, the spin quantum numbers m_{s1} and m_{s2} (Chapter 8) for the two electrons. Then

$$\Psi_a = \frac{1}{\sqrt{2}}[\Psi(x_1, x_2, m_{s1}, m_{s2}, t) - \Psi(x_2, x_1, m_{s2}, m_{s1}, t)] \tag{9.6}$$

is the type of wave function that will describe the state. Here you must understand by Ψ a function not only of the space and time variables but also of the spin variables—a function that assigns to each spin variable one of its two possible values, $+\frac{1}{2}$ or $-\frac{1}{2}$. The rule can be generalized to any number of identical particles.

The most useful application of the law of antisymmetric states is the *Pauli exclusion principle*: no two electrons can have all their quantum numbers, including spin, the same. Alternatively the principle can be restated: only two electrons, one with spin 'up' and one with spin 'down', can occupy 'the same state', where now the specification of 'the state' does not include a specification of the spin.

The relationship of the exclusion principle to the law of antisymmetry becomes clear when you notice an assumption implicit in the principle. 'The state' referred to in the principle is a *one-electron state*, not a state describing the whole assembly. The principle is assuming that you can describe the state of the assembly with good approximation by a combination of wave functions, each written for one of the electrons in the assembly.

The method by which this might be done is developed in the problems at the end of this chapter. For a system involving two electrons, the method might yield the approximate wave function

$$\Psi_a = \begin{vmatrix} u_k(x_1, m_{s1}, t) & u_l(x_1, m_{s1}, t) \\ u_k(x_2, m_{s2}, t) & u_l(x_2, m_{s2}, t) \end{vmatrix}, \tag{9.7}$$

where u_k and u_l are one-electron wave functions. Clearly this determinant is an antisymmetric function, for interchanging the subscripts 1 and 2 interchanges two rows of the determinant and so reverses its sign. Thus the function obeys the law of antisymmetric states. And clearly also, if the two electrons occupy the same one-electron state, $u_k = u_l$ and Ψ_a vanishes, giving for the two-electron state a probability of zero, in obedience with the Pauli exclusion principle.

PROBLEMS

9.1 Show by the method of separation of variables (Chapter 4) that if in a system of two electrons the electrons are uncoupled, so that

$$U_{pot}(x_1, x_2) = U(x_1) + U(x_2),$$

the solutions to the Schrödinger equation for the two-electron problem can be expressed as products of solutions for one-electron problems.

9.2 Show how such solutions as (9.7) above are constructed from the solutions of Problem 9.1.

9.3 Discuss the values of energy, corresponding to the two-electron wave functions of Problems 9.1 and 9.2, in terms of the energies of the component one-electron wave functions. How many distinguishable wave functions correspond to a given value of the total energy after you have taken into account the law of antisymmetry? In other words, how does the degeneracy of a two-electron state compare with the degeneracies of the component one-electron states?

9.4 Show that the definitions of a_0 and W_H in Discussion 9.1 give to those quantities the dimensions of length and of energy, respectively.

Index